WINDMILLS

Overleaf

. . . At eve thou loomest like a one-eyed giant
To some poor crazy knight, who pricks along
And sees thee wave in haze thy arms defiant,
And growl the burden of thy grinding song.

LORD DE TABLEY (1835–95)

Windmills

SUZANNE BEEDELL

Foreword by HENRY LONGHURST

DAVID & CHARLES
NEWTON ABBOT · LONDON · VANCOUVER

To

PETER DURHAM

whose speciality has nothing whatsoever to do with windmills
but without whose expert help this book
would never have been written

ISBN 0 7153 6811 7

set in Monophoto Imprint
and printed in Great Britain by Ebenezer Baylis Ltd Worcester
for David & Charles (Holdings) Limited South Devon House Newton Abbot Devon

Published in Canada
by Douglas David & Charles Limited 132 Philip Avenue North Vancouver BC

Contents

Foreword

I am much honoured to be invited to write a foreword to so erudite a book on windmills, for I am really no expert on the subject and have a most rudimentary idea of how they work, or used to work. On the other hand I have been the owner and am now, so to speak, the caretaker of what surely must be the best-known pair of mills in the country, namely Jack and Jill, which stand on the top of the South Downs overlooking Ditchling, Hassocks and Hurstpierpoint. As a matter of fact there is also the stump of a much older one called Duncton, built in 1774, and now a single round room of immense character, in which I am at this moment writing. So widely known are these mills that when, for a book on oil, I was visiting some of the remotest outposts in the Middle East and people asked the usual question, 'Where do you live?' and I replied, 'I am the proud owner of two celebrated windmills on the South Downs,' no fewer than eight different times people said, 'You don't mean Jack and Jill?'

Now I am no longer the owner, since the two local councils, appreciating that nowadays no one can maintain two disused windmills out of the remains of taxed income, took over their maintenance as landmarks in perpetuity; in return for this we must by now have had at least a thousand people, mostly children, coming in conducted parties to look over them. It gives us great pleasure that the mills in turn should be giving something back in this way.

Our house lies between the mills, so that, while we do not live in them – with the exception of Duncton and the old granary, perhaps one of the most beautiful rooms in Sussex, wonderful for Christmas parties and the like – we do live constantly *with* them, and they take on separate and individual identities matched only by railway engines in the, alas, departed days of steam. Jack, somewhat austere and obviously the male one, has 100,000 bricks and five storeys in ever-decreasing circles till you come to the cap, from which you may survey a 60-mile view ranging from the Devil's Punchbowl at Hindhead, 44 miles away, across to the North Downs and Reigate Hill, and thence to Ashdown Forest.

Jill, amazingly, also has five storeys and enough of the original machinery remains for even laymen like myself to gain a vague impression of how this mill and similar ones worked, of the dust and the flour, and particularly of the danger, with huge driving wheels turning, stones grinding, canvas belts on either side ready to take your arm off, and chutes here and there waiting to break your leg – as one nearly did mine – if you forgot exactly where they were. The driving wheels, all made of wood, really are a masterpiece of the carpenter's art – 8 ft in diameter, with innumerable wooden cogs let into the wood. By stepping outside his back door, high up at the top of the movable steps, the miller pulled a chain to rotate a wheel (both are still intact) and a ratchet and pinion device opened or closed the slats in the sails as they were actually going round; opening them was the only way he could stop them and make it safe enough to apply the wooden brake. This device was installed in 1832 and must have been the very first 'variable pitch propeller'.

You have only to climb up Jill's five precarious ladders from storey to storey to marvel at the skills of our ancestors and to reflect that we, with all our boasted technology, may yet come round to re-creating the likes of Jack and Jill to harness the wind for our energy once again.

HENRY LONGHURST

PLATE I

FULBOURN, CAMBRIDGESHIRE

Efforts are being made to get this now extremely derelict mill repaired. The weatherboarding is vertical on some quarters, horizontal on others.

Introduction

We live in a time of energy crises. Sources of power as we know them are running out. Except for a few nuclear power stations, and some nuclear submarines designed solely as lethal weapons, we have not come up with new forms of energy to run our motor cars, our homes, our factories and our farms. It could be that before long we shall be back again in the position of primitive man looking for sources of power, and wondering how to use and control the enormous forces of the winds that blow continuously across the surface of our planet. For a few centuries, wind power was used to drive ships and to turn windmills geared to various forms of machinery.

Nowadays, windmills are at work only in underdeveloped countries, or preserved for the sake of the past and because they are beautiful and interesting, not because they are practical economic propositions. The problem is the same as it has always been – that wind power is not constant. Except in a very few places the wind cannot be relied upon to blow steadily, so the windmill is an unreliable machine which cannot be left to work by itself for more than a few minutes at a time. To give of its best, it must have a human attendant who never leaves it and is prepared to put it to work whenever the wind blows, day or night, high day or holiday.

Small modern mills which pump water from below ground, or which turn small generators, can be left to themselves, but their product is not constant, and wind pumps usually have a motor pump to fill in for them when there is no wind. Mills which generate electricity are useless when the wind fails, and have back-up systems of batteries to store extra power; but these systems, if they rely solely on wind, are far too haphazard to be of much practical use, except in remote areas where there is no alternative.

As steam, oil, and electric motors, reliable and constant, were developed, windmills were abandoned. When things fall into disuse or go out of fashion, they tend to be forgotten for a couple of generations and then to re-emerge – not as practical things but as antiques, curios, or simply objects of beauty. The artifacts of our immediate predecessors do not appeal to us, so most of them are destroyed or ignored until they decay. A few survive, either hidden or still in use in out-of-the-way places. Then one day these survivors are rediscovered, to become valuable and treasured. So it is with windmills. Most have gone, both here and in Europe, but just before it was too late to save them, many were renovated or preserved. This work is still going on.

I am old enough to have seen working windmills in England. The first thing I ever remember could well be a windmill! I was staying with my grandmother at Fulbourn in Cambridgeshire; she left there when I was three years old, so I must have been about two and a half. One of my aunts took me for a walk in a push-chair, up a path, across a field, to a windmill standing at the top. I remember being frightened by the shadow of the sails as they turned fast in a fresh breeze, and awed by the descending 'swoosh' of each sweep as it passed by. (They must have been common sails of canvas – patent sails are almost silent.) The miller came out and stood by the doorway. We went inside and I remember the gentle rumbling noise of the mill at work. A beam of sunlight coming in through a small window was full of dancing motes of flour dust, and the miller himself was covered and fringed with the same dust which made him seem a strange kind of man to me.

Fulbourn mill is derelict now, even more derelict than it is in the photograph (plate 1), as it lost a sweep some years ago when it was struck by lightning. It has a new owner who is very interested in repairing the mill, but nowadays it costs thousands of pounds to restore what was originally built for a few hundred.

Fulbourn mill was built in 1807 by Mr John Chaplin. There is still in existence his copy of *Old Moore's Almanack* for that year, in which he wrote down some details of his mill. He laid the first brick on 11 July and began grinding on 15 December. Five months to build a mill does not seem to be very long, especially as the work must

PLATE 2
HIGHS MILL, POTTER HEIGHAM
NORFOLK
A fine Norfolk drainage mill with an elegant tower and a boat-shaped cap with petticoat. Since converted into a home.

PLATES 3 AND 3A
HORSEY MILL, NORFOLK
As it is now and as it was about 1914. This mill has had a chequered career. The original structure, built in the early eighteenth century of tarred brick, was known as Horsey Black mill. It was tailwinded in 1895 and lost its cap. The tower was rebuilt early this century with red bricks from a kiln at nearby Martham and, apart from a new vertical shaft of Scandinavian pine, the original machinery was re-used. In 1943, the mill, still pumping, was struck by lightning which split the sail stocks. Since then it has been repaired by Norfolk County Council and SPAB, and now belongs to the National Trust. From the cap gallery, there is a fine view across Horsey Mere to Hickling. The mill has a traditional boat-shaped cap with vertically boarded petticoat, but now with neither shutters nor striking gear, which are clearly visible in the old photograph.

have gone on through the autumn with its equinoctial gales and into the early winter. The cold winds from the north-east sweep across the wide open flat lands of East Anglia and the low hills of Cambridgeshire; then, as now, it could have been no fun working on the outside of the mill, or out on the sails at that time of the year.

My family later moved to Norfolk, where I was really in windmill country, drainage mills this time. I can remember when you could stand anywhere in Broadland and see half a dozen mills twirling away over the tree tops. These drainage mills were replaced by motor pumps in the 1930s and almost all of them have now disappeared or are in ruins. Only a handful have been preserved to enhance the scenery. For years after that I hardly saw a windmill, and in fact never saw one working for its living until five years ago when I went on holiday to Holland. But they fascinate me now more than ever, and always have.

About ten miles from my home is Chillenden mill, the last post mill to be built in Kent. She stands alone and lonely on the top of a low hill, where once you could have looked around and

PLATE 4
CHILLENDEN MILL, KENT
One of the last open post mills to be built – in 1868 – it stands on the site of a much older mill shown on a map dated 1695. The present mill worked until the end of World War II grinding animal feed, but was not well balanced and the whole structure canted forwards until a trench had to be cut in the ground to allow the sweeps to go round. Kent County Council repaired the mill as a memorial to old millwrights. Now fixed to face the prevailing wind, it has fixed skeleton shutter frames, where once were patent shutters. The elliptical springs and tensioners at the end of each sail remain. The mill is once again in need of repair.

seen half a dozen other working mills near and far. She has been restored and fixed to face the prevailing wind. But this year her paint is peeling again; half a sweep has fallen rotting to the ground, and her tail wheel is crumbling out of its iron tyre. Again hundreds of pounds will be needed to put her right. Although it takes me through lanes and a roundabout way, I visit her as often as I can. Chillenden mill will never turn again and stands, like too many others, motionless while the wind forever passes her by.

Hopefully, more and more people are becoming interested in windmills, and there are young men working as millwrights again. Local and national societies are taking up the cause and there is an ever-growing body of *aficionados*. I shall try in this book to put together technical and general information about windmills both in Britain and abroad, so that anyone interested can learn enough about them to be able to study and enjoy them, with an understanding of how they work and how they differ. I shall also illustrate how windmills were part of the everyday life of our ancestors – and completely taken for granted.

1

Origins and history

When early man ceased to be a hunter and settled down to grow grain, he had to find a way to turn it into flour. To crush it on a stone by beating it with a piece of wood or with another stone must have been the first method; this was followed by rubbing it between two stones, one stationary and one turned by hand. Next came horse mills turned by harnessing an animal to a beam and making it walk in a circle. The need for more constant sources of power to turn the heavy stones and speed up production soon brought about the invention of simple watermills, and over the centuries water power was harnessed to ever more sophisticated designs of mill. But where water was not so plentiful – in the Middle East, for instance – another source of power was necessary. There is evidence that windmills existed in Persia in the tenth century, and were possibly known some three hundred years earlier – an Arab geographer of the seventh century mentions windmills in use in Persia.

Mechanically the Persian windmill was a development of the watermill, and looked a little like one turned up on end. A vertical shaft was set in a square tower, with sails set at right angles to the shaft at its top end. The wind passed through slots in the tower on to the sails and thus turned the shaft and any machinery set at the bottom end. Traditionally this principle travelled with the prisoners of Ghengis Khan to China, where horizontal mills with matting sails, but without the enclosing tower, were used to drive irrigation machinery. The Chinese discovered that if the sails were 'feathered' – set at an angle to the shaft – the mill would turn continuously so that there was no need for the enclosing wall and slots to direct the wind on to the driving side of each sail as it passed the slot.

In England, the earliest written record of a windmill is a rental note dated 1185 of a windmill in a village called Weedly, in Yorkshire, let for eight shillings a year. In 1191 there is a record of a mill at Bishopstone, in Sussex. This leads us to believe that windmills must have been invented earlier in the twelfth century. Whether they appeared first in England or in Europe, nobody knows, but either way their construction and use would have spread rapidly throughout windy Northern Europe. In France there is an oblique reference to a windmill in a deed c 1180, which gives a piece of land 'near a windmill' to an abbey in Normandy, and reference is made in *The Chronicles of Jocelyn de Brakeland* to a windmill built in 1191 at Bury St Edmunds. In Holland, the earliest known record of a windmill relates to a privilege granted to the burghers of the town of Haarlem by the Count Floris V in 1274 – a lot later than the Weedly record.

By this time there was a papal ruling on tithes from windmills so they must have been pretty general, although it was another hundred years before they became common in southern Europe, and two hundred years before they appeared in Spain and Asia Minor. Perhaps the truth is that the invention of the windmill occurred quite separately and in various forms in all those different places. The Middle Eastern windmill has always been very distinctive from the northern type, and the Spanish variety resembled the Middle Eastern and North African windmill rather than that of northern Europe.

The earliest known illustration of an English windmill is in the *Windmill Psalter* (c 1270), originating in Canterbury, but now in the Pierpont Library in New York. This and many other manuscript illustrations and later carvings – notably the sixteenth-century carving on a misericord in Bristol Cathedral – show simple post mills; but in France tower mills were more common and there is a French manuscript in the same museum showing a tower mill of a type very similar to that still existing in north-west France.

Smock mills were a much later development. They probably first appeared towards the end of the seventeenth century, both in Britain and in Holland, but some authorities believe they existed as early as the fifteenth century. The great Dutch polder mills, although thatched rather than weatherboarded, had in principle the same basic wooden framework as the English smock mill, and

PLATE 5
This painting, dated 1642, by the Dutch artist van Goyen, shows a typical open post trestle mill of the Low Countries, on high piers, with a long ladder. It is of the same basic construction as the post mill at Coquelles, in France (plate 89), built in the late 1700s and still working.

first appeared in the 1600s. In Britain, brick and stone tower mills mostly appeared later. All three types continued to be built until the windmill ceased to exist as a working machine.

A mill can be defined as a 'building fitted with machinery in which any of various mechanical operations or forms of manufacture are carried out'. Obviously, once windpower was harnessed through the sails and machinery of a windmill, it could by various ingenious devices be made to do all kinds of jobs. In Britain water power had already been put to a lot of industrial uses and, for work other than grinding corn, windmills were nothing like so commonly used as in Holland where they did many different jobs. To turn a watermill, running water is needed which can be dammed up sufficiently to give a constant head. There was plenty of fast running water in hilly Britain, but hardly any in Holland. In such low-lying country, there was too much water, which had to be pumped uphill by windmills into dykes and rivers flowing slowly to the sea. As you will know if you have ever visited it, Holland is a windy country, where the wind can be relied upon

PLATE 6
DUTCH POLDER MILL
With thick thatch and decorated beard behind the stocks, this is a fine example of a type of mill found in many areas of Holland.

to provide plenty of mechanical power.

Sawmills were developed first in Holland at the end of the sixteenth century, and it was not long before the first sawmill was built in England. There were many fulling mills and quite a few wind-powered sawmills, some of which remained in use well into the present century. Windmills were used for grinding whiting, ochre, snuff, cement clinker – Berney Arms mill ground cement clinker which was dug up at Burgh Castle and carried to the mill in wherries – bark for tanning, white lead, flint for china and logwood for dye. In addition to these, rice was hulled, and cocoa, pepper and mustard were ground in Dutch mills.

Drainage mills and irrigation mills – where the power of the wind was used to lift water from one place to another – were in fact developed before grinding mills in some countries, such as Persia and China. In Holland particularly they came into use after 1400, as soon as sea defences and dams were built so that pools and lakes could be drained; in increasing numbers, and ever more efficient types, they continued to be used until engine power replaced them.

Generally speaking the windmill in Britain and in northern Europe lasted for 700 years. To medieval man it must have seemed a strange creature, a squat beast with whirling arms, enough to terrify the wits out of any wandering countryman unfortunate enough to come across one of the new-fangled things. Very soon familiarity would remove fear, and it was a poor parish indeed that did not have access to at least one mill.

When wynd wantyd at the wyndmyll he carryd the corne to the watermyllne, and when water was scant he carryed the corne to the wyndmyll.

Harleian MSS, British Museum

PLATE 7
BERNEY ARMS, NORFOLK
This big tower mill served a dual purpose: it was used for grinding clinker carried to it by wherry and as a drainage mill. The detached scoop wheel, boat-shaped cap and gallery, and the stage high up outside the tower from which the striking gear could be worked, show clearly. This mill can only be reached by boat or by train.

As much part and parcel of community life as the church, the mill supported and was supported by a series of craftsmen: the millers themselves; millwrights who built and repaired mills; *stone dressers* who tended the millstones; sawyers and carpenters who provided the timbers; masons who built the towers, and later iron founders who made cast-iron parts enabling many technical developments to be made.

By the beginning of the nineteenth century there were thousands of windmills. In 1768 there were twenty-seven marked on the map of Liverpool. Except in the heart of the great moors and forests and mountains, wherever there was a view extending more than a few hundred yards, you could see windmills, bringing life and movement to the whole landscape. Nothing nowadays replaces that sight.

Then steam power made large-scale roller milling possible under factory conditions in towns and in ports where grain was imported. Windmills were steadily put out of business by the development of reliable engines, by transport which could take grain to big mills and move flour out to bakers and grocers, and finally by World War I, when flour-milling standards were set by law. Windmills could not reach the flour-milling requirements, and grinding purely for animal feed was not so profitable. When the mills needed repair, the expense was not worth while; in a vicious circle, millwrights were forced to charge more for their work as costs generally rose, so there was less and less work for them, and fewer and fewer men carried on the trade. By the end of World War II, millwrighting was almost a forgotten art, many mills had vanished for ever, while many more were derelict, with only a handful in working order. Now, thirty years later, with active societies and thousands of individuals interested in windmill preservation and quite a few millwrights working, many mills in Britain are no longer tragic eyesores.

If you want to see more than one windmill working at the same time, go to Holland, where in 1923 they woke up to the fact that they were losing their wonderful windmills and did something about it; they have now saved nearly a thousand. Wherever you happen to be, visit a windmill which is not just turning for show but is doing the job for which it was built. Savour the warm smell of crushed grain and come out powdered with the fine creamy dust of fresh

Margate Mills.

PLATE 8
'The three mills here represented, are objects of considerable attraction, and from standing on very high ground, are seen at a considerable distance by sea and land: they are situate near Dane or Hooper's Hill, where once stood the horizontal mill invented by Captain Hooper, and called by his name. This mill was pulled down a few years ago, and is now remembered only as an ingenious piece of machinery; and on account of the following curious occurrence, which, while it proves the extraordinary violence, and the irresistible power of the wind, is deserving of relation from it's singularity. During the dreadful gale by which the old pier was destroyed, the top of this mill, with the upper tier of sweeps, fliers, &c., weighing more than four tons, was torn from it's iron fastenings; and after being blown a distance of several hundred feet, was lodged in a field near the present mills, without sustaining or doing any mischief.'
– from an old book about Margate

flour. If you have been in there long enough it will be on your clothes, on your hair and eyebrows, and in your mouth! Or stand by the water wheel of a drainage mill as it rhythmically splashes and scoops its thousands of gallons. Then you will come to understand something of the power of the wind, and to regret the loss of so many windmills.

Milling Soke

Before the Norman Conquest many freemen held watermills, but after 1066 – when it was realised how economically important they were – mills, watermills, horse mills and hand mills were all governed by what was known as *milling soke*. Feudalism was monopolistic within its manors and, although milling soke was never the subject of an Act of Parliament and therefore never repealed, it was part of the charter of each manor, and as such was enforceable. In fact, it gave the lord of the manor – whether he were king, church, monastery, baron, or even squire – a monopoly over milling. Every mill was the property of the lord of the manor on whose land it stood. Even if a man dared to build a mill of his own, it was judged to be the property of the lord of manor, unless, as happened more and more as time went by, the mill was let to a miller with its rights of soke. Eventually, of course, mills were built in disregard of milling soke, but up until the last century there were continual legal battles to enforce soke rights, and mills were still being sold together with soke rights, rather as one sells the goodwill of a business today.

In my grandfather Edward Moore's time there was two or three proud fellows set up mills, but he preferred a bill in the Duchy showing how our windmill (Townsend) is the King's mill, and the tenants within Liverpool ought to grind there, because he paid a great rent. And after two hearings it was decreed that those new erected mills should be pulled down and fined besides; which was accordingly put in execution, and the mills were pulled down.

My grandfather, as he was tenant of your King's mill (Townsend), preferred his bill in the Duchy against . . . both of whom erected horse mills in the town; after a great suit there was a decree made that as he was the King's farmer within the King's manor there ought to be no private mills, and those who erected them were fined and both the mills pulled down.

Church Society Publication, *Moore Rental* (1847)

On the face of it the various soke laws sound reasonably fair, but because most of the rural

PLATE 9
WESTGATE MILL, KENT
Standing by the seashore, this Kentish smock mill, typical of its period, has a wheel and chain winding gear operated from ground level.

population were dependent upon lords of the manor or their agents for almost everything, and could find things made very hard for them if they did not toe the line, milling soke may in fact have been a very restricting monopoly. 'Monopoly' is a dirty word to most of us in the twentieth century, yet milling soke should not be condemned out of hand just because it was monopolistic. It enforced the availability of mills, and it stabilised prices on a national basis. In medieval times lack of transport and roads prevented grain or flour being carted over any but short distances. The lord of the manor was supposed to provide and maintain enough mills to meet the demands of his people. A Dutch source reckons that one mill could grind sufficient grain for 2,000 people, but this seems very optimistic. Certainly in their heyday there were many more mills per head of the population.

If the lord of the manor demolished a mill or let one fall into disuse or disrepair without doing something about it immediately, the soke was broken and his people could take their grain elsewhere. On the other hand he was able to insist on rigidly laid down rules for grinding – which varied little throughout the country: 1/16th of the grist for home-grown corn and 1/24th for bought corn. Within these limits there was some variation and until the thirteenth century 1/20th was common toll. Payments then began to be made alternatively in money, and in 1796 money payment was enforced by statute.

In France definite limits were prescribed beyond which tenants were not forced to travel to the manor mill. In Poitu, Touraine and elsewhere, the limit was 2,000 strides of 3 ft each from the tenant's door to the mill. In Anjou and Maine the limit was 1,000 revolutions of a wheel 5 ft in circumference; in Brittany it was 60 cords of 60 ft each. This works out at just under a mile and a quarter in the first instance; just under a mile in the second, and around three-quarters of a mile in the third – not far at all, even when humping a sack on your back. And there would have to be mills within $2\frac{1}{2}$ miles, $1\frac{3}{4}$ miles and $1\frac{1}{2}$ miles of each other in these areas.

The lord of the manor could demand that his own grain was ground free (free toll); in some cases, even if the hopper was full of someone else's grain when his arrived, he could demand that the hopper be emptied and his work attended to at once (hopper free).

> And when the lord's corn come to the miln he shall put all men out of their grist, and take their corn out of the hopper if there be any therein; and his corn shall be ground next, before all men, when it comes to the miln, without multer or paying service to the milner, but as his lyst (if he likes) and curtasy to come to the said miln.
>
> Church Society Publication (1422)

Those who did not grow grain, but bought it inside the manor, had to have it ground at the manor mill; if they tried to take it elsewhere, they would more often than not get caught and find themselves in trouble. Grain bought outside the manor could usually be taken anywhere for grinding, but some charters demanded that it should be brought back into the manor. This was to prevent farmers from selling their grain outside the manor to its own residents, to avoid the milling soke restrictions.

Eventually it all became too much trouble. A lord of the manor with huge estates, such as the king, let off his mills to tenants, and enforcement became almost impossible. As mills became old fashioned and techniques improved, they fell into disrepair and this freed the residents to use other mills. The rows that took place over milling soke in the eighteenth and nineteenth centuries resulted from the efforts of those who had bought mills and paid good money for the milling soke rights trying to enforce what had become an unenforceable monopoly.

2
How windmills work

To state the obvious, when the wind blows upon the sails of a windmill, they turn, and thus rotate the axle, called a *windshaft*, upon which they are set. Inside the top of the mill, also on the windshaft, is the *brake wheel*, a big wheel with cogs on it which, by means of gearing, drives the millstones and all the subsidiary machinery in the mill. In drainage mills the drive is carried down to the base of the mill and out to turn a big *scoop wheel* (plates 107, 108).

Before the wind can strike the sails at the correct angle to make them turn, they must be brought to face the direction from which it is blowing. The whole of the body (of a post mill) or the cap (of a tower or smock mill) is turned – or *winded* – by various means.

The *stocks* form a cross on the windshaft and carry the *whips* which bear the framework of the sails. The 'disc' described by the revolving whips is exactly at right angles to the wind, and the bars mortised through them – which carry the *cloth sails* or the *shutters*, as the case may be – are set at an angle to this imaginary disc.

As the wind strikes the angled sail, some of its force (called the tangential force) is exerted to push the sail away from it, so the sail turns. The greater the sail area (within limits) the stronger this force will be and the faster the sails, and consequently the windshaft and machinery, will turn. But sails turning too fast create enormous stresses of thrust and friction, which a mill could not sustain indefinitely; in any case, while a mill needs to go at a fair speed (at least twelve to fifteen revolutions of the windshaft per minute) in order to grind or pump efficiently, nothing is gained by it going too fast. Mills with more than four sails are certainly more powerful, but the weight, stress and expense factors probably outweigh the advantages.

Sails

New ways of operating windmills were evolved during the Middle Ages. These came about by trial and error or by accident – empirically – rather than by design. Those semi-literate millers did not sit down at a drawing-board or make mathematical calculations; they observed, and tried things out. With a watermill the principles were obvious and simple. Moving water pushed things in front of it, so it would push the paddles of a wheel set vertically or horizontally in it, thus turning the axle of the wheel. Wind also pushed things in front of it. It may have been simple enough to devise a small windmill on a tripod, using flat sails like long paddles. Eventually it would have been found that the best angle at which to fix the sails on the stocks would be somewhere about 20° from the plane of the stocks. The earliest sails therefore were set up with a *constant pitch* of about 20°. They consisted of a simple wooden cross mortised through an axle or windshaft, with *bars* mortised through the stocks and sticking out at an equal distance on each side of them. Later the ends of the bars were mortised or bolted into longitudinal *hemlaths* to brace the whole sail. Some of the older types of Continental mills still carry bars set up without hemlaths (plate 86).

Cloth sails were laced in and out of the bars, one each side of the stock and tied on at both ends. The sails were reefed by moving them in towards the middle like curtains. When bigger windmills were built, it was realised that stocks long and strong enough to take the whole sail were not easily available, and separate whips were made to carry the bars. These whips were bolted and clamped to the stocks and had several advantages. If one sail broke, it could easily be replaced without disturbing its opposite number, or without having to be taken out of the windshaft. The mortise in the end of the windshaft was always a weak point, because water could get into it causing rot and eventual breakage at this point of great strain. The invention of cast iron solved this problem. A big casting, known as a *poll end* or *canister*, was fitted to the end of the windshaft, the stocks were fitted into this and the whole thing was secured and tightened up with wedges or clamps.

North of a line from the Severn to the Wash,

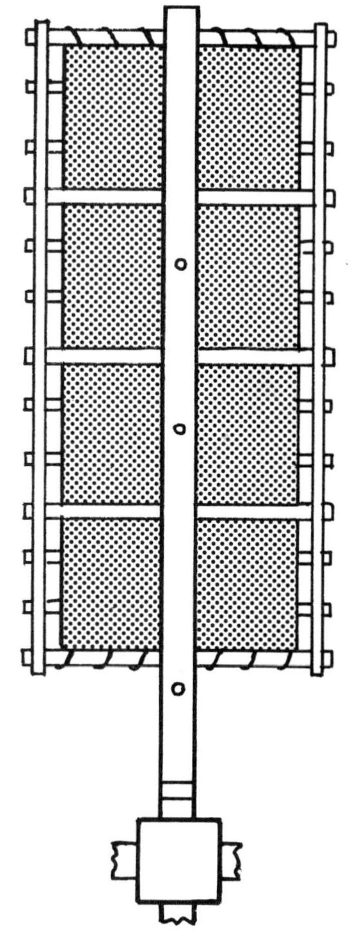

FIGURE 1 Cloth sails laced.

PLATE 10
EASTRY MILL, KENT
WINDSHAFT AND BRAKE WHEEL
The windshaft passes out of the front of the mill over the neck bearing on a block resting on the breast beam. The metal band of the brake passes round the outside edge of the wheel. The cogs have been removed from the brake wheel.

PLATE 12
CANISTER AND WINDSHAFT FROM BATTLE MILL, SUSSEX

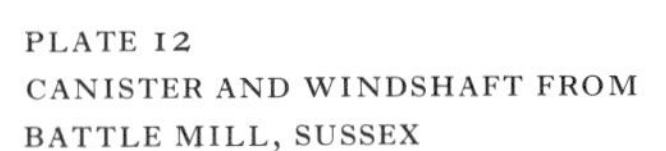

Now on display at Polegate mill, Sussex, this is about as close a view as one can get of this part of a mill. The knob on the front was very useful as a point of attachment for pulleys when fitting new stocks or sails.

PLATE II

BLACKDOWN MILL, PUNNETTS TOWN, SUSSEX

The stocks pass through the canister with wedges holding them tight. The whips are clamped and bolted to the stocks, with side clamps bolted through to strengthen the whole assembly. The make-up of the sail bars is also very clear.

particularly in Lincolnshire, cast iron provided another means of fixing stock and sails to windshaft. An *iron cross* was attached to the windshaft, and sail backs or heavier whips were bolted directly to this, without stocks being used at all. The advantages of this method were that sails were independent of each other and that water would not lie up between cross and sail back, so the junction would stay even drier than by using the canister.

The iron-cross poll fixing made it possible to put on five, six, or even eight sails, because it was easier to make a cross with that many arms than to make complicated canisters. Windmills with more than four sails became common in the east Midlands, and there were a few in other parts of the country. The idea was that the additional sail area increased power, and the extra sails made the mill run more smoothly. There was a safety limit to the width or length of sails because of stresses and weight; this restriction prevented the four sails being made wider or longer, although in some Norfolk mills they were as much as 10 ft wide. Five-sailed mills became very unbalanced when a sail broke, but multi-sailed mills with an even number of sails could continue to work if one broke, although it was necessary to remove the opposite one of a pair to adjust the balance (plate 102).

PLATE 13
PILLING MILL, LANCASHIRE
IRON CROSS POLL END
The sails were individually bolted to this. The derelict cap frame and curb can also be seen.

The next development in design was the common sail which, with variations of detail, is still in use. The bars protrude less on the sail's leading edge than on the trailing edge, and in many cases there are only longitudinal boards, *leading boards*, in front of the whip (plate 16a). The cloth sails are tied at the inner end or *heel* of the sail frame or attached to rings on an iron rail. The sail has lines or cords running right down both edges, and pointing lines attached at intervals lower down the trailing edge, so that the sail can be rolled and *reefed*. This cuts off the outer lower corner of the sail, thus reducing area and the power developed. There are four settings, first reef, sword point, dagger point, and full sail.

In order to start the mill, each sail in turn is brought down so that the cloth can be unrolled and spread across the bars. In a light wind the whole cloth is spread, but should the wind get up, or be too strong for the full cloth, then the mill must be stopped by braking and the cloth reefed accordingly. It is absolutely essential that the miller's judgement is good, for if too much sail is set and the mill goes too fast it can be extremely difficult to stop, and too much pressure on the *brake* can cause enormous friction and perhaps set the mill on fire. If the brake cannot control the sails, the mill will not stop and anything can happen; the sails turn much too fast and 'run away', and the resulting stresses and strains may do all kinds of damage. A sail may break up and catch on the body of the mill, or behind other sails, wrecking the whole windmill; but fire from the sparks caused by friction is the greatest danger. Once the mill is stopped, the miller must go down and reef the sails; there are accidents on record caused by the brake slipping and the mill beginning to turn while the miller was still taking in his reefs.

As time passed, millwrights – influenced by experiments carried out by John Smeaton – discovered that *weathered sails*, which are angled at about 20° at the inner end and flattening to about 5° at the tip, seemed to be more efficient. The bars are mortised through the whips at the appropriate angles right down the sail, and *backstays* are fitted to strengthen the sail and to maintain the pitch of the bars (plate 11). The angle is

PLATE 14

BARDNEY MILL, LINCOLNSHIRE

An old photograph of a six-sailed mill with an ogee cap. The crinolines date the picture.

PLATE 15

WICKEN FEN DRAINAGE MILL CAMBRIDGESHIRE

This little mill is on a nature reserve; it works effectively and, although tiny, has some interesting traditional features. The sails are attached at the inner end like curtains on rings. Winding is by tailpole and braces from the boat-shaped cap, but this cap has vertical weatherboarding.

PLATE 16
'MICHEL', COQUELLES, CALAIS
SETTING COMMON SAILS
(a) the miller unties the sail cloth. Note the wooden leader board directing wind BEHIND the sail, like a jib, to enhance 'lift'; (b) flicking the rolled up sail cloth off the cleat; (c) pulling the sail cloth across the bars with the pointing lines; (d) tying down the spread sail.

(a)

(b)

(c)

(d)

PLATE 17 (opposite)
DUTCH WIP MILL IN FULL SAIL
With full sail set, this mill is going well and the water is welling out of the tail race. Wind about force 4–5.

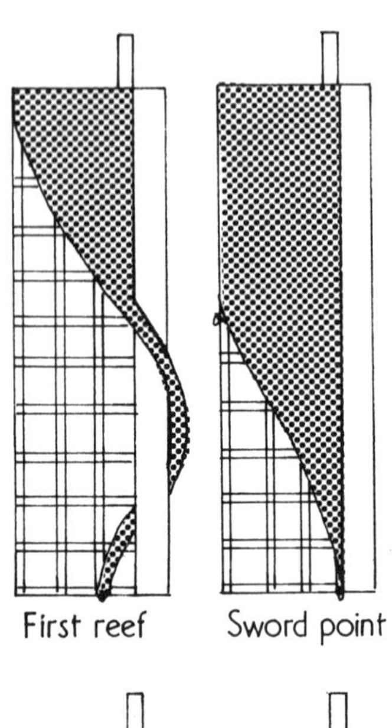

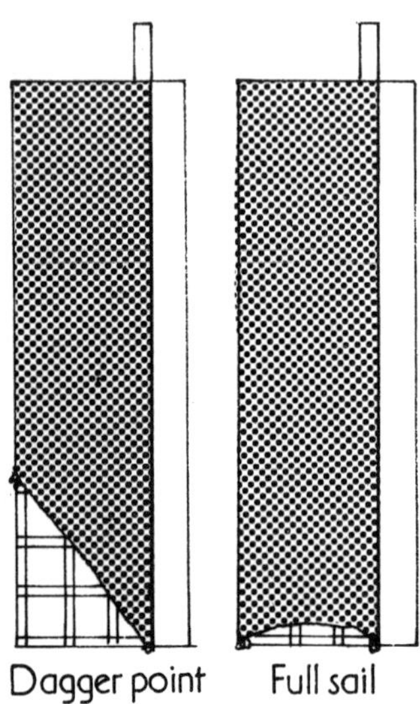

FIGURE 2 Sail reefs.

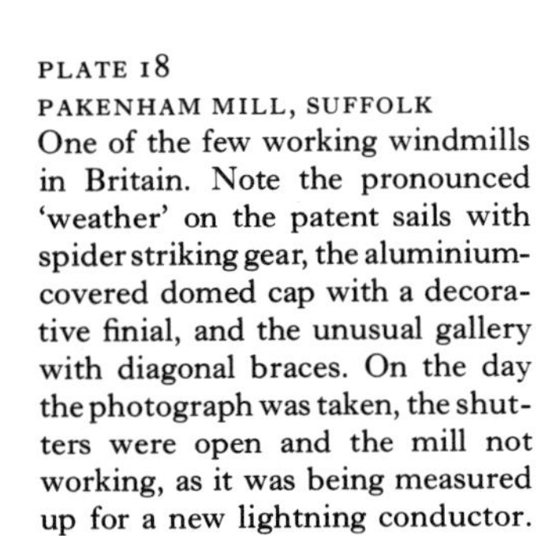

PLATE 18
PAKENHAM MILL, SUFFOLK
One of the few working windmills in Britain. Note the pronounced 'weather' on the patent sails with spider striking gear, the aluminium-covered domed cap with a decorative finial, and the unusual gallery with diagonal braces. On the day the photograph was taken, the shutters were open and the mill not working, as it was being measured up for a new lightning conductor.

greatest near the centre and smallest near the tip of the sail, so that the sail has *varying pitch*, ie the angle between the plane of the circle in which it rotates lessens from centre (20°) to tip (5°). This varying pitch is supposed to increase the aerodynamic efficiency of the sail, but there is a school of thought which believes that the pitch can remain nearly constant along the whole length of the sail and be almost as efficient. Weathered sails are slightly harder to start than constant pitch sails.

In parts of the north of England, and in Lincolnshire particularly, millwrights fashioned their sails with flicked-up top outer edges to the downwind side, rather like the turned-up edge of a sheet of paper. This sail shape is common in Holland and was possibly copied from there.

In 1772 a Scottish millwright called Andrew Meikle invented the *spring sail*, which went into common use, especially in England, and is still seen on many mills (plate 4). Hinged shutters of wood or painted canvas on a light wood frame, each about 12in wide, are set into the framework of the sail, so that when the shutters are closed, the sail offers a flat surface to the wind. The shutters are linked by iron cranks to a wooden shutter bar running the length of the sail. By moving this bar up and down along the sail, the shutters open and close like the slats of a venetian blind. The shutter bars are attached via an iron rod to elliptical leaf or coil springs at the inner end of the sail. These springs can be tightened or loosened from the outer end of each sail by means of a rack and pinion or a simple tensioner (plate 4); the miller sets each sail in turn according to how he judges the average strength of the wind. Then, when a strong gust comes, the pressure on the shutters overcomes the tension of the spring and the shutters open, spilling wind and slowing the sail. As the wind dies away, the tension of the spring overcomes the pressure of the wind and the shutters close, presenting a greater surface to the wind and maintaining the speed of the sail.

This is an excellent system, although the mill has to be stopped if alterations need to be made to the tensions of the springs. Nevertheless, the much greater latitude allowed by the *spring-loaded shutters* means that the mill can be kept working with fewer interruptions. Spring sails are not so efficient as cloth common sails, although they will work within a 10° angle each side of the wind. Mills sometimes carry a pair of each, a compromise which allows at least some self-regulation.

The next development was obvious and inevitable – an arrangement whereby the setting of the shutters could be altered without stopping the mill, and which linked all the sails together. The springs were abandoned and, instead, each sail carries a metal rod, called a *fork iron*, which runs halfway down the sail where it is attached to the shutter bars linking the cranks on the individual shutters. At the inner end these irons are each attached to a *bell crank lever* or triangle, joined in turn via short links to the *spider* or cross on the end of a *central striking rod*, which runs right down through the hollow windshaft. The striking rod passes out of the end of the windshaft to a position either just inside or outside the mill, where it is coupled to a *rack member*. Meshing with the rack is a pinion wheel with a *chain wheel* mounted on the same shaft. An endless chain hangs over this wheel, and pulling on it causes the striking rod to move up or down, pushing or pulling the spider and the bell cranks, thus opening or closing the shutters. As an alternative to the rack and pinion and chain wheel, a large *rocking lever* attached to a bell crank may be linked to the striking rod.

To start the mill, the chain is hauled till the shutters are closed, the brake having been taken off inside the mill. Then, when the mill is turning at the required speed, the miller hangs a weight on the chain. As the wind slackens, the weight pulls the chain down and closes the shutters; as the wind gusts, it overcomes the weight, lifts it slightly, and the shutters open spilling wind. So the mill turns more or less steadily, with the weight moving quietly up and down, and the shutters opening and closing all the time. To stop the mill, the weight is taken off and the chain hauled to open the shutters and spill all the wind. Then the brake is put on steadily and the mill stopped and held.

Some of these so-called *patent sails* are fitted with *air brakes*. These are two longitudinal shutters in the leading edge of the tip of the sail. When the sail turns too fast, these turn outwards – actuated by the same mechanism as the shutters – break up the air flow and slow down the sail.

Jib sails are common in Holland. The Dutch corn mill in plate 20 has two variable pitch jib sails and two common sails. A sailing dinghy can

PLATE 19
BLACKDOWN MILL, PUNNETTS TOWN, SUSSEX
A smock mill on a wide brick base. The sails have little or no weather, almost constant pitch.

PLATE 20

COLLINSPLATZ MILL, HOLLAND

A brick tower mill, with a thatched cap, at work on two patent jib sails grinding corn for animal feed. Note the flicked-up end of the common sail and its furled canvas, and the decorated beard. Sail bars are set directly into the stocks.

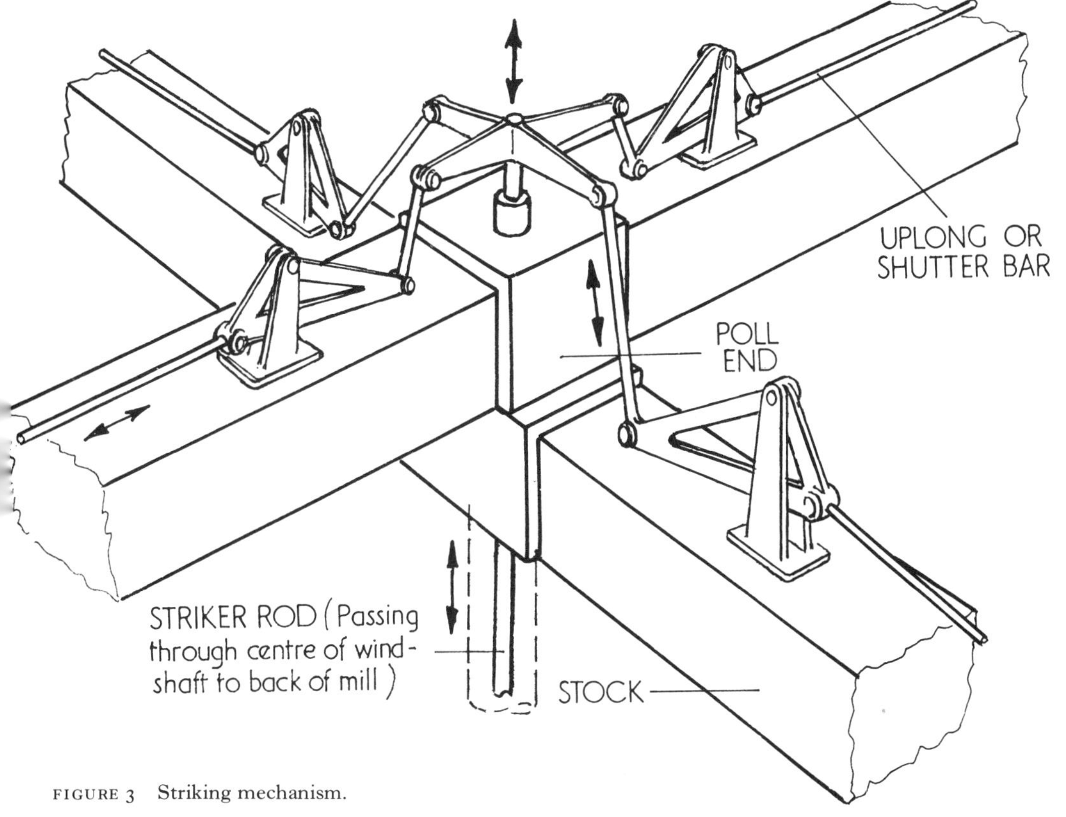

FIGURE 3 Striking mechanism.

PLATE 21

STELLING MINNIS MILL, KENT

A shutter, covered with white-leaded canvas.

PLATE 22
CHILLENDEN MILL, KENT
The elliptical springs which control the shutters. Note also the iron canister carrying the stocks, the whips bolted on, and the access hatch in the front of the mill.

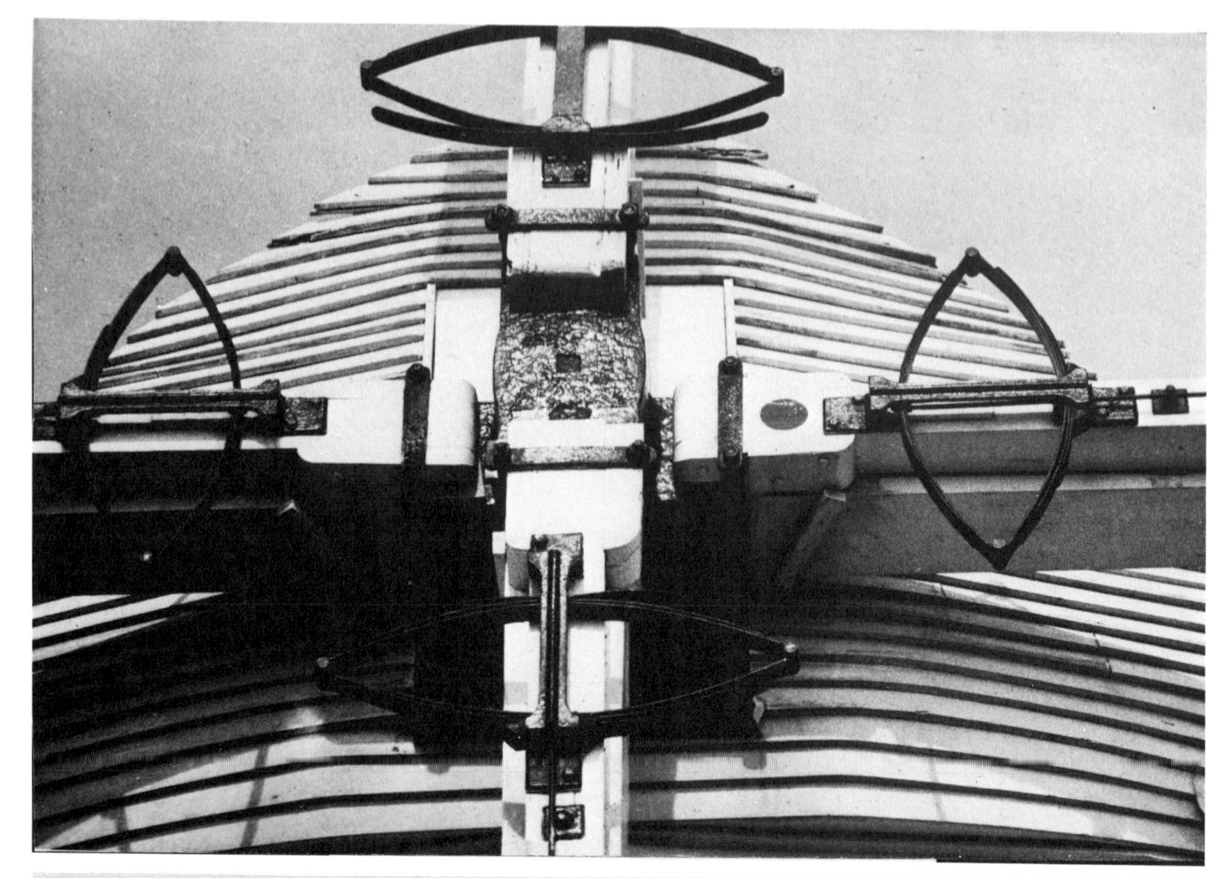

PLATE 23
STELLING MINNIS MILL, KENT
A worm's eye view of the spider striking gear. The mill has only one pair of sails on and the striking rod can be seen passing through the empty section of the canister into the windshaft.

PLATE 24
HERNE MILL, KENT
The rocking lever beneath the fantail stage with the endless chain attached. A weight on the lower end balances the pressure of the wind on the shutters and the rod actuates the striking mechanism accordingly.

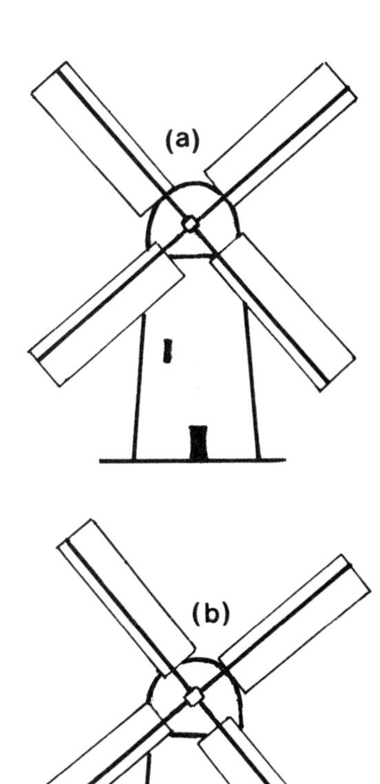

FIGURE 4 (a) clockwise and (b) anticlockwise sails.

move forward in the lightest puff of wind because the jib sail directs the breeze along the BACK of the mainsail. The mainsail has a curve which creates an incomplete vacuum on its lee side. This incomplete vacuum causes the boat to be sucked forward as the air tries to fill it. The breeze from the jib enhances this effect. Nowadays the jib sails of a mill are commonly made out of metal with streamlined shapes like aeroplane wings, the leading edge being attached to the whip so that it directs the wind BEHIND the trailing edge. These will turn in the very lightest of breezes. If jib sails turn too fast problems arise, so they usually have an air brake fitted in the leader board.

Sails turn clockwise or anti-clockwise according to which way the bars are set on their whips. When the mill was stopped, the miller took great care to leave his sails tidily set. He could judge to a nicety by watching the sails pass a window or door exactly when to put on the brake to achieve the required position – usually that of a St Andrew's cross – in which the sails were well balanced and sustained the least strain. When at rest, the sails could be set to convey a particular meaning. In England, sails left in the St Andrew's cross position signified that the mill was shut down for some time and the miller not present. Sails left in the St George's cross position indicated only a brief stoppage, or that the miller would soon return – a kind of 'back in ten minutes' sign. Other positions could indicate that the services of a stone dresser were required. These settings meant different things in different places. In Holland a mill with the upper sail before the vertical position signified a celebration; with the upper sail past the vertical signified mourning.

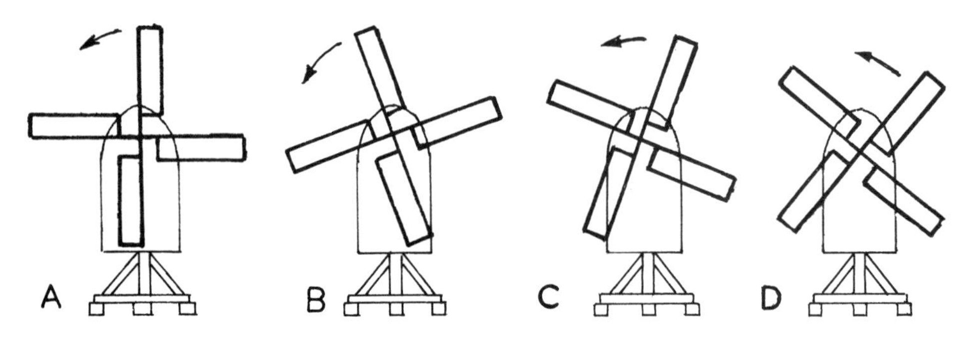

FIGURE 5 Position of sails; A: St George's Cross, 'At home' or 'Back soon'; B: 'Mourning'; C: 'Celebrating'; D: 'Packed up' or 'Away for the day'.

In England shutters were sometimes removed from sails in mourning, the number taken out being in direct proportion to the closeness of the relationship of the deceased to the miller. One can imagine that this led to all sorts of family rows, when relatives felt slighted because the miller had not taken out enough shutters!

There were up to 10,000 mills in England of which a high proportion were windmills, these were so prominent that obviously they could be used for signalling, although there could not have been much secrecy about it. Nevertheless there are tales that, in areas where smuggling went on, windmills were frequently used to convey warning signals. On a prearranged code, the sails would quietly be turned to a certain position at the approach of the revenue men, for instance, thus passing a message very quickly over miles of countryside. At night a lantern could be tied to the tip of a sail and hoisted to a prearranged point.

Drive Machinery

Very early in the history of the windmill, the drive mechanisms were invented and perfected, and the principles have remained the same, except for some changes from wood to iron and steel. The early gear wheels were made of wood with inserted wooden teeth or round pegs – these teeth being made of hard woods, often apple or hornbeam. Smaller wheels often consisted of two discs with wooden staves between, called *lantern pinions*. As time passed, gearing became more refined and eventually cast-iron gearing was introduced (plate 27). Wooden teeth meshing with iron was used a lot in later mills, as it was found to be quiet and lasted well. An iron wheel with wooden cogs is called a *mortise wheel*.

The huge brake wheel set in the top of the mill, with the windshaft as its axle, drives everything else. The brake wheel is so called because it has a brake round it, either an iron band or a wooden shoe (plates 25, 96). By moving a lever, this brake is made to grip the wheel and slow it down, and also to hold the sails in one position while the miller is working on them. In light winds the miller will use the brake lever to stop the mill; to avoid the necessity for climbing up to work the brake lever, if he is setting sails at ground level outside, he has a line passing right down with which he can work the brake (plate 88). To attempt to stop a mill with the brake alone, if it is really travelling or has 'run away', can be fatal (as already explained the friction causes showers of sparks which can set the mill alight), although with common sails the miller has little option but to use the brake. In a post mill, by choking the stones with grain and attempting to *quarter* the mill by turning it manually away from the wind, might succeed in slowing it to the point where the brake could be effective, but this would be impossible without help and very difficult with it. Patent sails which spill wind make the use of the brake much safer.

Some post mills drive only one pair of millstones directly from the brake wheel, which is a face or bevel wheel. The bevelled lantern pinion *wallower* – a wheel set horizontally – in plate 25 is axled directly on to the *quant* which runs down and drives stones set immediately below it. At the other end of the windshaft in this mill is another wheel with a lantern pinion behind it driving another pair of stones below. This is known as a *head and tail mill*.

In spur gear post mills, two pairs of stones are set side by side in the head and are driven in the same way as in tower and smock mills (see below).

Tower and smock mills take their drive from the brake wheel by means of a wallower. Its vertical axle, the *upright shaft*, runs down to another big wheel at its bottom end, the *great spur wheel*.

Smaller spur wheels, called *stone nuts*, take the drive from the great spur wheel. In plate 28 the drive is DOWN by means of the quant to the runner stone 'overdrift'; in plate 29 the drive is UP by means of the spindle to the runner stone 'underdrift'. These stone nuts can be moved to put them out of gear.

The various sack hoists, flour dressers, grindstones, and other subsidiary machinery, are all run from the main drive using cog or friction wheels or belt drives which can be brought in and out of gear very simply (plate 61). Some can be done while the mill is working. The miller did not reckon to do much heaving when he had all that wind power available. He hoisted his grain to the top of the mill and it fed by gravity through the grinding machinery until it reached the flour bins at the bottom.

In drainage and pumping mills there is much less machinery – only the brake wheel and wallower at the top, and a bevel *crown wheel* instead of the spur wheel at the bottom of the upright shaft. This crown wheel drives a much bigger

PLATE 25
'MICHEL', COQUELLES, CALAIS
The brake wheel of a Flemish post mill, with a lantern pinion wallower directly driving a pair of stones immediately below it. Note the wear on the well-greased staves of the pinion, also the massive wooden brake block round the wheel. The pinion wheel on the right is on an axle which passes to the back of the brake wheel where a pulley carries a belt; this can be seen passing down to the flour-dressing machine (out of the picture). Overdrift.

PLATE 26
PAKENHAM MILL, SUFFOLK
Modern gearing.

FIGURE 6 Principles of the brake.

FIGURE 7a Gearing and stone layout of post mill with two underdriven sets of stones in the breast.
FIGURE 7b Gearing and stone layout of 'head and tail' post mill with two overdriven sets of stones.
FIGURE 7c Gearing and stone layout of post mill with a single stone in the breast.

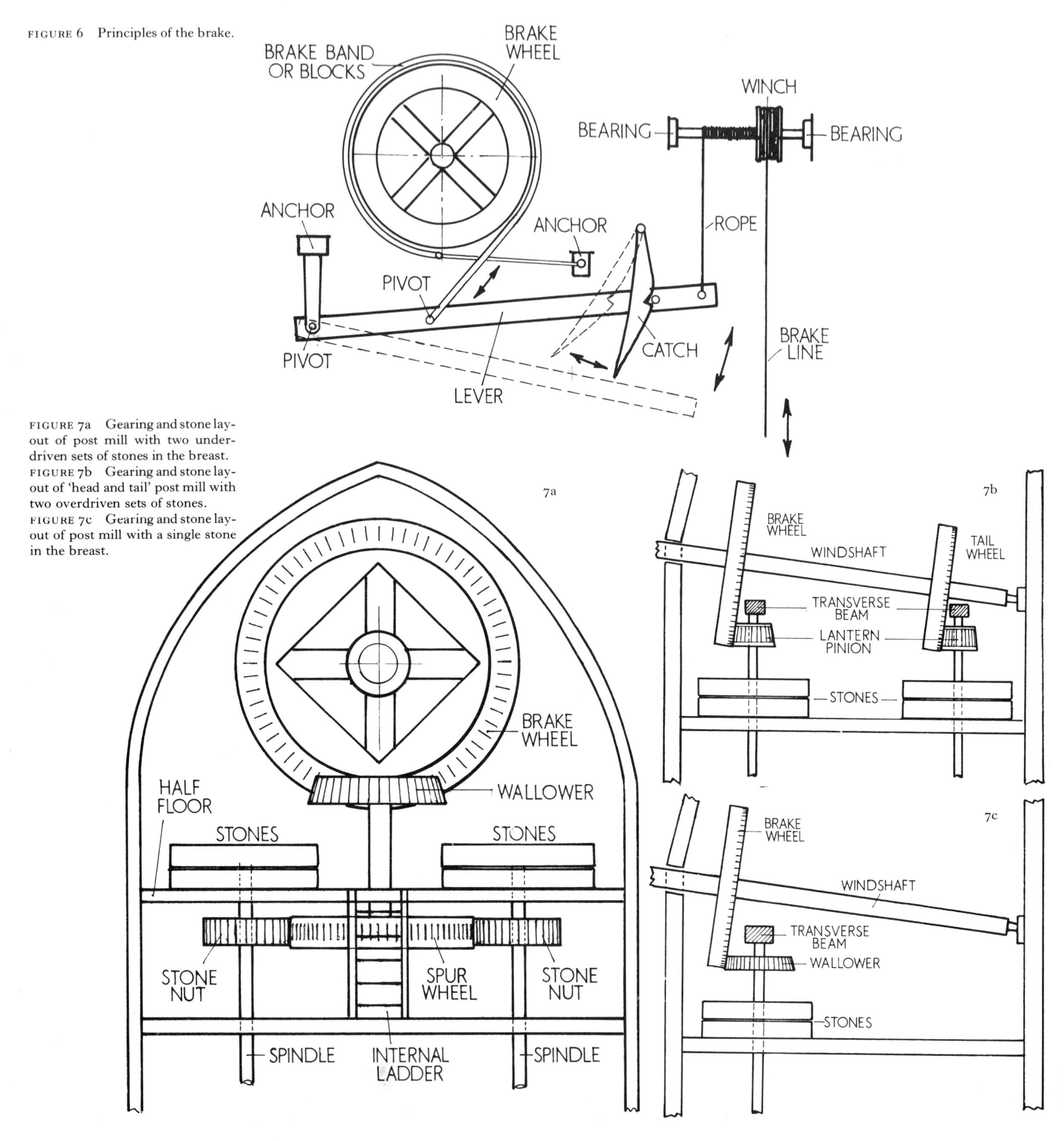

PLATE 27
UNION MILL, CRANBROOK, KENT
The great spur wheel with a bevel wheel below it. The stone nut on this side of the wheel has gone, but the quant can be seen beyond the wheel coming down from a stone nut on that side. Overdrift.

PLATE 28 (below right)
PAKENHAM MILL, SUFFOLK
The great spur wheel with stone nut and quant. Meal spouts to the hoppers below. Overdrift.

PLATE 29 (below left)
STELLING MINNIS MILL, KENT
Great spur wheel and stone nut, which is put in and out of gear by the screw below it. Underdrift.

FIGURE 8 Underdriven and overdriven stones.

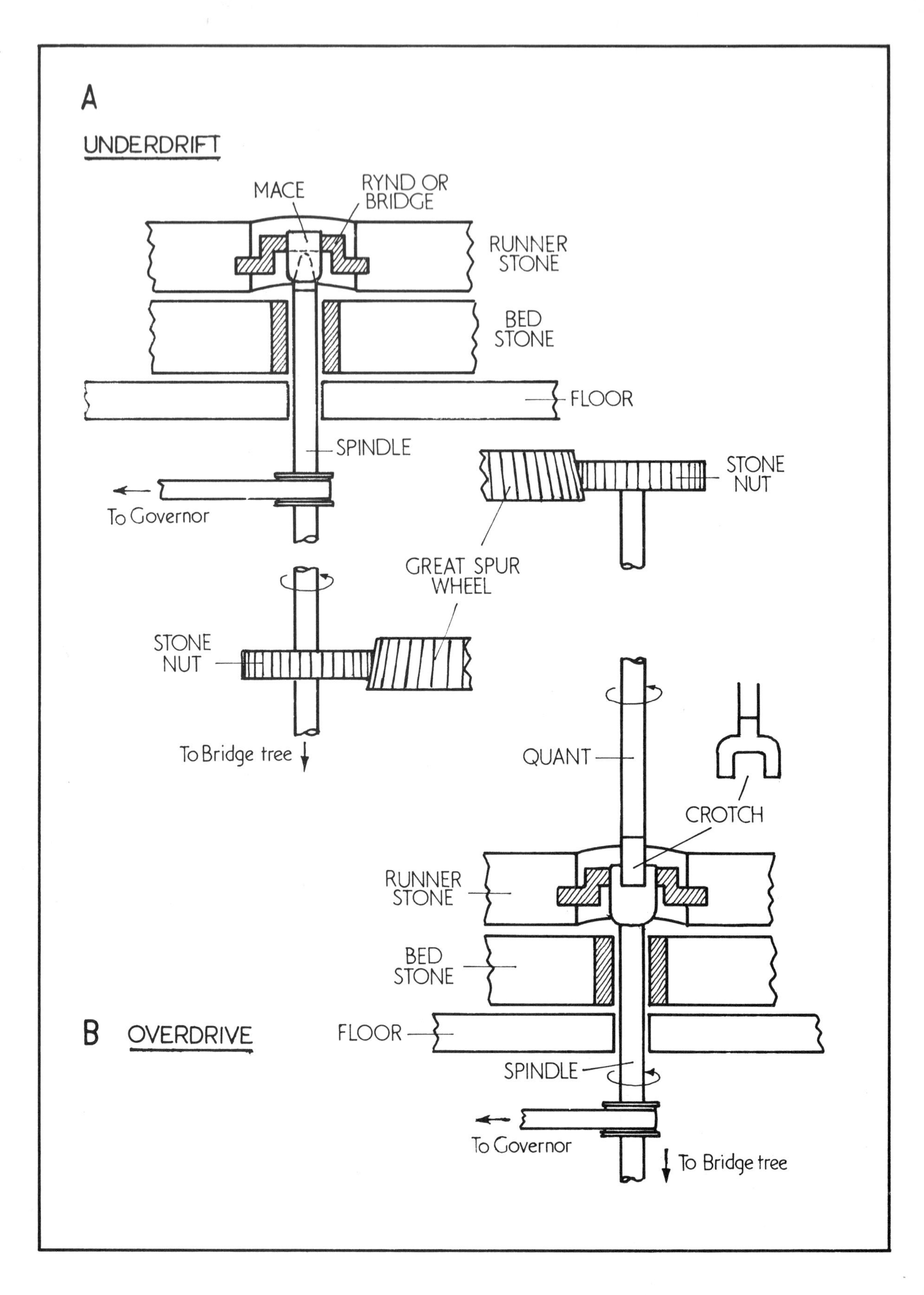

PLATE 30
'MICHEL', COQUELLES, CALAIS
The exterior sack hoist. Notice the weather vane, small cross, and decorative barge boards even on the little roof over the hoist. This hoist runs from gearing on the tail wheel just inside the mill.

PLATE 31
CHEDGRAVE DETACHED MILL
NORFOLK
An old drainage mill, with tail beam and braces. The tail beam support at ground level is still there, and the ribbing of the boat-shaped cap shows the pleasing curves of this design.

pit wheel on the horizontal axle. This axle passes out through the side of the mill to turn the scoop wheel (plate 108). Other types of drainage mills had a centrifugal pump on a vertical shaft.

Tailpoles and Fantails

Early post mills were winded by hand, the long beam or *tailpole* attached to the body of the mill being pushed round by brute force. Tower and smock mills winded by beams and tailpole from the cap became rarer in England, although quite a few East Anglian drainage mills worked this way. The little mill at Wicken Fen (plate 103) still has its beams and braces, and Chedgrave detached mill was an example of the type in Broadland.

On the Continent, this method of winding has persisted with little variation. Dutch 'wip' mills (hollow post mills) have a capstan wheel at the bottom of the ladder, and the huge polder mills, which are smock mills with caps, have beams coming right down from the cap with a capstan wheel at the bottom (plate 32). This wheel has a chain wound round its axle. On the end of the chain is a loop, and right round the mill are a series of little bollards. The loop is dropped over a bollard and the wheel is turned, winding up the chain on the axle so that the tailpole moves towards the bollard, pulling the body of the mill round.

The French post mill called Michel has a windlass at the bottom of the tailpole. This is wound with a crank and the wire is hooked to rings set in the ground.

The Spanish pumping mill shows very clearly indeed the simplest form of tailpole coming down to ground level from the back of the movable cap (plate 110).

Chillenden mill (plate 44) has a tailpole with a wheel at its end, exactly the same in principle as that on the old mill at Irby (plate 46). This supports the end of the tailpole and the wheel runs on a paved or iron track.

The mechanical *fantail*, which in one form or another was developed in Britain, seems to be so utterly logical that one wonders why it was not universally adopted on the Continent. I found when visiting Holland that the millers I met were more interested in my photographs of fantails than anything else.

Basically, a fantail – or *fly tackle* as it is called in some areas – is a secondary windmill set at right angles to the main sails of the mill. It is small and has several *vanes*, so that it spins fast. It is geared through to a worm or a spur pinion so that it turns the cap of the mill on its *curb*, which has rack teeth. As the wind shifts to one side the fly begins to spin, and the cap turns bringing the main sweeps to face the wind. The fantail then being edge on to the wind ceases to turn until the wind changes again, however slightly.

If you watch a fantail you will notice that it

PLATE 32 (inner left)
WINDING CAPSTAN ON A DUTCH MILL
The capstan is attached to the bottom of the tailpole and ladder, high above ground; note the chain passing from windlass to bollard, and the anchor chain. A man can get a lot of strength to bear pulling down on the capstan arms at this level.

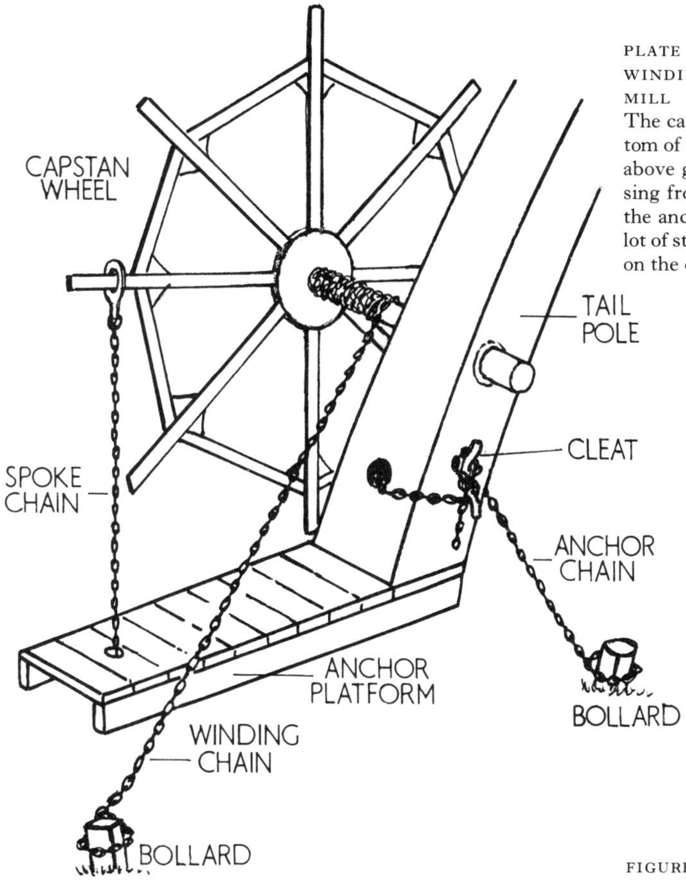

FIGURE 9 Capstan.

constantly makes a few turns one way or the other, and that the cap moves almost imperceptibly in response to it. To turn the cap through 360° may take as long as five minutes, but the response is normally quick enough for the safety of the mill. It is only in extremely rare turbulent and violent stormy conditions that the wind changes too quickly for the fantail to turn the cap. When that happens the mill can be *backwinded*. The wind striking the sails from behind causes the shutters to close with a bang and the solid sail area this presents to the wind creates enormous stress on the cap, which will probably come right off the curb, smashing the sweeps against the ground.

Here is a vivid description by Mr J. Bryant, the miller of Pakenham mill (plate 18), of what happened one stormy day in 1947:

It was about 3.30 in the afternoon and I didn't like the look of the sky. The mill had been working with the wind in the south-east, but the wind dropped right away. I struck up the mill at once and chained the striking bar to the fly post. Whilst I was doing this there came a little air from the north . . . the fly ran, and got back as far as north-east and stopped. By this time I could see a great cloud approaching from the west accompanied by sheet lightning and rolling thunder.

PLATE 33
'MICHEL', COQUELLES, CALAIS
This winding gear works like a winch, with a cranking handle, and ratchet and pawl gear wheel. The wire winds round a drum made of flat pieces of wood set on a hub, and hooks to iron rings anchored firmly into the ground.

PLATE 34
STELLING MINNIS, KENT
FANTAIL VANE
The axle and casting which holds the six-bladed fantail. The boards are set at an angle on their stocks to achieve pitch.

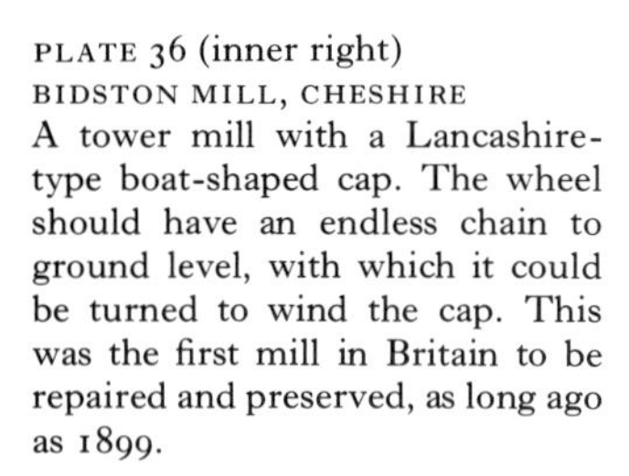

PLATE 36 (inner right)
BIDSTON MILL, CHESHIRE
A tower mill with a Lancashire-type boat-shaped cap. The wheel should have an endless chain to ground level, with which it could be turned to wind the cap. This was the first mill in Britain to be repaired and preserved, as long ago as 1899.

As I was up with the fly I had a good view over the surrounding countryside and could see the direction and speed of the approaching storm.

I hastily turned the fly by hand, and had nearly got the mill back to north when the gale hit her. I moved inside the mill very quickly . . . I had never seen a fly go so fast, nor have I since. She raced into the storm and went dead west, then two minutes later went back to the north.

The storm eased later and the wind returned to the west and began to lull, but had I not turned the fly by hand as I did she would have been tail winded and the top of the mill would have been blown off. This storm was the worst I ever experienced.

Before the invention of the fantail, many mills had a rack attached to the curb, geared to a wheel set at the side or back of the cap. An endless chain hung over this wheel, right down to the ground, and the cap was winded by pulling on the chain and thus turning the wheel (plates 9, 36). Then the fantail was invented to make the winding automatic. The fantail can be put out of gear so that the mill can be winded by hand, and where the drive is through a worm, the *chain wheel* described above has been retained for this purpose on a few mills.

Post mills are normally winded from ground level, as the whole body of the mill goes round. To make this automatic the tailpole is mounted on a tandem carriage, with a big fantail above it. The fantail drive runs down to the wheels and as the wind turns the fantail, so the carriage moves one way or the other on a hard circular track round the mill, taking the body with it.

In East Anglia, the fantail is fixed directly to the ladder, not to the tailpole (plates 38, 47). This is a stronger and more stable arrangement. The fantail on the tailpole may act as a weathervane and swing the mill without turning. The fact that both these types of fantail are low down, and therefore screened to some extent from the wind by nearby buildings, trees and the mill itself, might make it less effective than a cap fantail. However, this loss of efficiency is counterbalanced by the fact that the patent sails will work (although

PLATE 35

PAKENHAM MILL, SUFFOLK

The eight-bladed fantail on its staging. A more sophisticated design than plate 34, with an iron hoop to brace the vanes and bolts between the vanes at the tips to maintain pitch.

PLATE 37
CROSS-IN-HAND MILL, SUSSEX
A ground-level fantail on a tandem carriage running on a hard track. The rods and gearing which drive the wheels can be clearly seen. The bottom of the ladder also runs on a track and has wheels to support it.

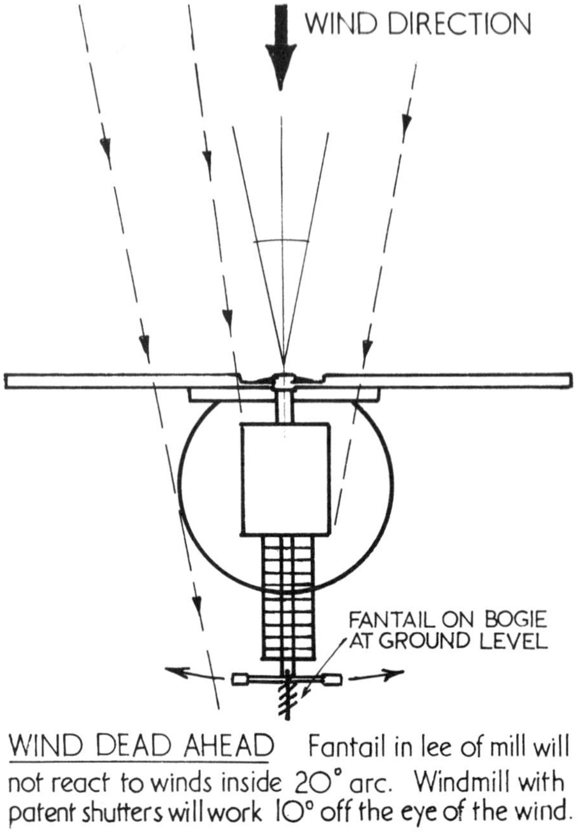

FIGURE 10 (inner right) Ground-level fantails.

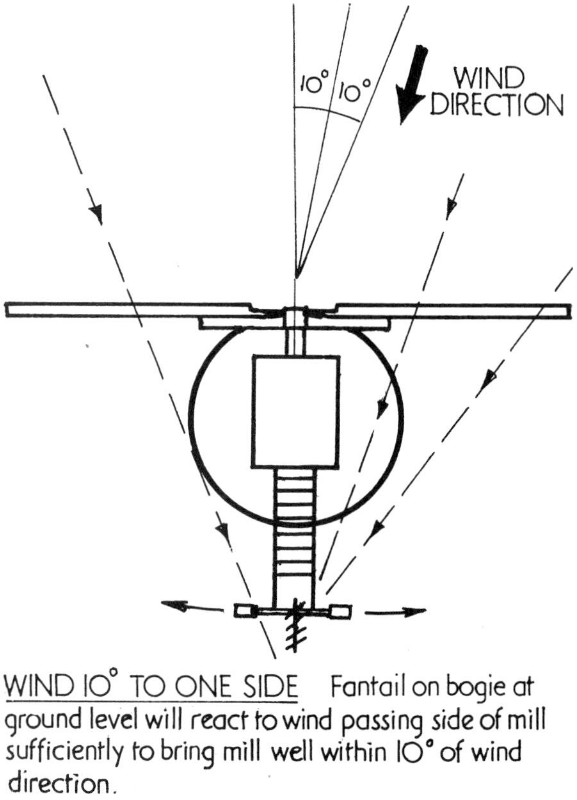

WIND DIRECTION

FANTAIL ON BOGIE AT GROUND LEVEL

WIND DEAD AHEAD Fantail in lee of mill will not react to winds inside 20° arc. Windmill with patent shutters will work 10° off the eye of the wind.

10° 10°

WIND DIRECTION

WIND 10° TO ONE SIDE Fantail on bogie at ground level will react to wind passing side of mill sufficiently to bring mill well within 10° of wind direction.

PLATE 38
HOLTON MILL, SUFFOLK
A six-bladed ground-level fantail attached to the bottom of the ladder, not to the tailpole. Note the hard circular track on which the carriage runs.

PLATE 39
HOGG HILL MILL, ICKLESHAM
SUSSEX
A post mill with a roof-top fantail.

PLATE 40
CLEY MILL, NORFOLK
Banded fantail, cap and stage.

not all at full power) provided the mill is within 10° of the wind, giving a total working arc of 20°. The fantail will always catch enough wind for it to turn and bring the mill into a 10° working arc (much better than the 20° maximum), so it is doing its job. In practice the ladder fantail does turn the mill dead into the wind in spite of being sheltered, and it would never have become so common had it not done so, because there is always a big drop in power if a mill is not right in the eye of the wind, no matter what the type of sail.

Some post mills were winded by a fantail on the roof driving down to a worm ring on the post just below the body, others had a fantail drive from the roof to wheels at the bottom of the ladder. These arrangements were not common because they were not very successful and the former tended to put too much strain on the gearing.

Usually fantails were painted white, but some millers decorated them with coloured bands. In Suffolk there was a fashion for alternate red, white and blue vanes.

3

English post mills

The mills shown in the earliest illustrations, dated 1270, are all post mills. The little mill at Bourn, in Cambridgeshire, looks very like one of these, but in fact it was built in 1636, though perhaps a little earlier. Although it has been altered and repaired many times, it retains its medieval shape. Drinkstone post mill in Suffolk, dated 1689, does not look like a medieval mill as it has been altered and rebuilt so much. Chillenden mill in Kent (plate 4) was built in 1868, but few post mills were built after the eighteenth century.

The span of 600 years is a very long time for any particular man-made object to have remained more or less constant in general type. If you could bring a miller from the reign of Edward I back to life in the England of today, there is not much he would recognise – some country churches and cottages perhaps, and the shape of a ploughshare. But show him a post mill and he would know instantly what it was, how it worked, and what it was for. In fact he could probably work it, once he was shown how patent sails behaved, although he might be very puzzled by the fact that no one seemed to be using the mill!

The body of a post mill – known in East Anglia as the *buck* – which carries the sails and contains all the machinery, is mounted on an upright main post, usually of oak, about 20 ft long, on which it can turn through 360° so that it can always be made to face the wind. On a *pintle* on the top of the post pivots the *crown tree*, a heavy oak beam high up in the buck. The crown tree is usually set forward of the centre of the buck so that the weight of the sails is balanced out. The part of the buck in front of the crown tree is called the *breast* or *head*, that behind it is the *tail*. As the mill design developed, *iron gudgeons* and iron socket and bearings were used instead of the oak pintle. The whole of the body of the mill is built up on the crown tree and has to be strongly made with mortise and tenon joints, and properly balanced so that the *bedstones* always remain level no matter which way the mill is turned. Some post mills have a collar on the post, where it enters the base of the buck, on which some of the weight is carried.

The foot of the post is mortised round two massive horizontal beams called *cross trees*, the ends of which rest on masonry or brick piers; it does not rest directly on the cross trees, merely being held in position by them. To spread the stresses and weight, to brace the post and keep the whole thing upright, there are four diagonal *quarter bars* which run from the waist of the post, just below the buck, to the top of the cross trees just above the piers. Occasionally there were three cross trees and six quarter bars. Some Continental mills have double or even treble quarter bars. The whole of this arrangement below the buck is known as the *trestle*. Chillenden mill and others like it, where you can see the trestle, are 'open post mills'. The pictures and carvings of very early mills show the quarter bars going down to ground level and not resting on piers, and it is probable that there were foundations or cross trees below ground level. Lifting the timbers off the ground on piers prevents rotting.

In early mills the windshaft passed horizontally into the top of the mill and carried the brake wheel, which turned a wallower on a vertical shaft which went straight down to drive a pair of millstones. It was soon realised that the horizontal windshaft wore its wooden bearings very quickly so that it actually began to angle downwards; then, as it turned, sails and windshaft would tend to screw themselves out of the mill and wreck the whole thing. Also, any forward angling of the sails could cause them to catch on the buck as they revolved, with disastrous consequences. A windshaft angled upwards 5°–15° from the horizontal is far better, and was soon developed to become standard. The *thrust* of the shaft is then taken on a *bearing* at its tail end, the sails are tilted well clear of the body, and the general balance of the mill is much improved.

At the back of the mill a wide ladder with hand rails gives access to the door. This ladder is hinged at the top so that it can be lifted clear of the ground when the mill is turned or winded.

PLATE 41 (opposite page)
BOURN MILL, CAMBRIDGESHIRE
Britain's oldest post mill, which is known to have been there since 1636. It retains the medieval shape, the roof being steeply pitched. Much altered inside as time has passed, it has one pair of common sails and one pair of spring sails without shutters or springs.

PLATE 42
DRINKSTONE MILL, SUFFOLK
This very old post mill was working, when this photograph was taken, on those two dilapidated sails. It has a brick and flint roundhouse, and the buck has been rebuilt several times.

FIGURE 11 Post mill basic design; a – Chillenden post mill, Kent: (a) main post; (b) quarter bars; (c) cross tree; (d) crown tree; (e) windshaft; (f) middling or stock; (g) whip; (h) brake wheel; (i) wallower; (j) upright shaft; (k) millstones; (l) sack hoist, chain driven from pulley on windshaft; (m) brake lever; (n) belt drive pulley for dresser (not shown) driven from bevel ring on great spur wheel; (o) tailpole; (p) steps; (q) talthur; (r) great spur wheel; (s) jack screw for lifting stone nut (obscured by great spur wheel); (t) spring; (u) pivot; (v) collar;
b – Post mill: the buck (A) pivots on the trestle (B). The crown tree (C) rests on a pintle (D) on the top of the post (E), which is held in position by the cross trees (F) and braced by the quarterbars (G). The whole structure rests on the piers (H).

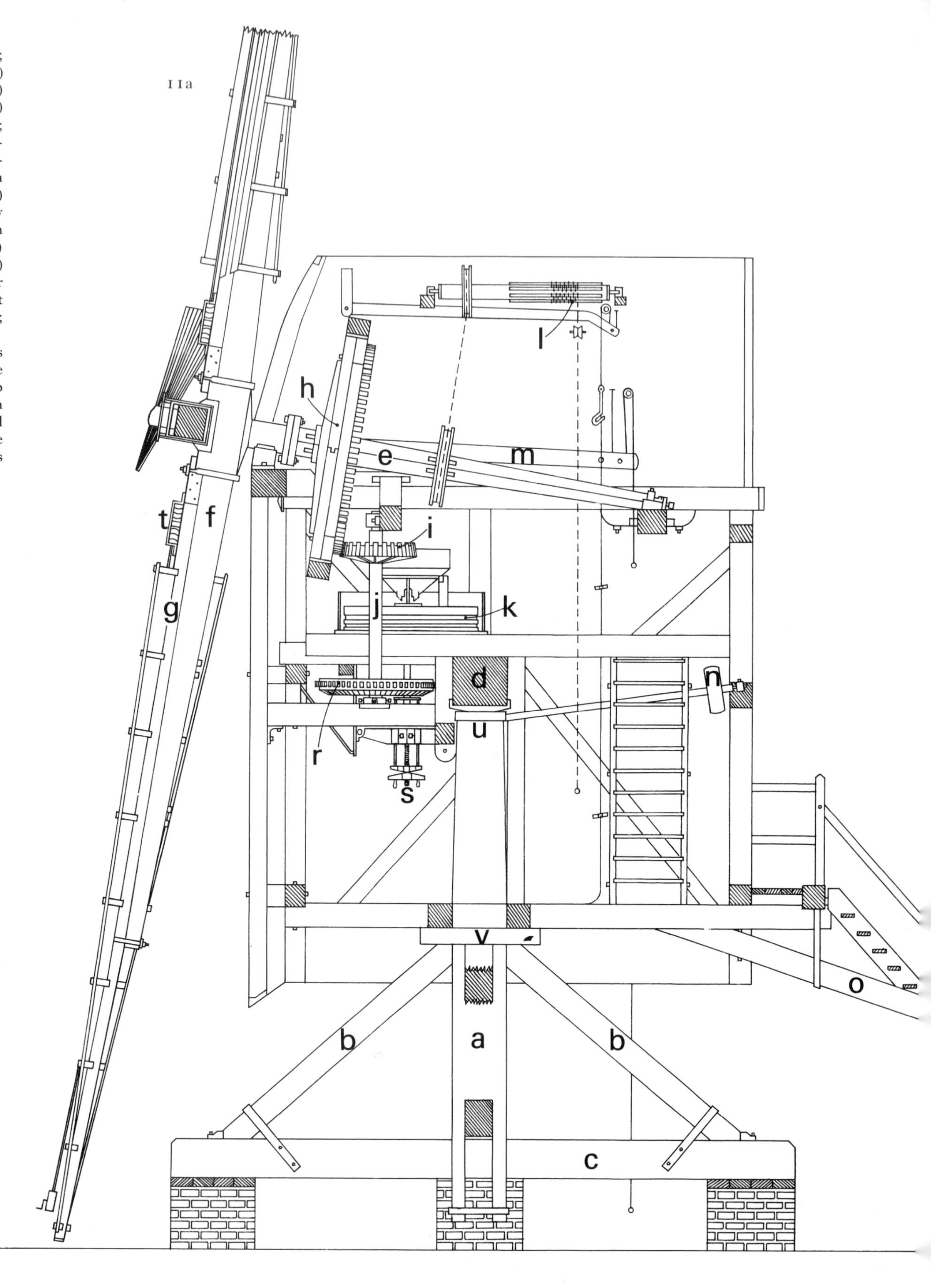

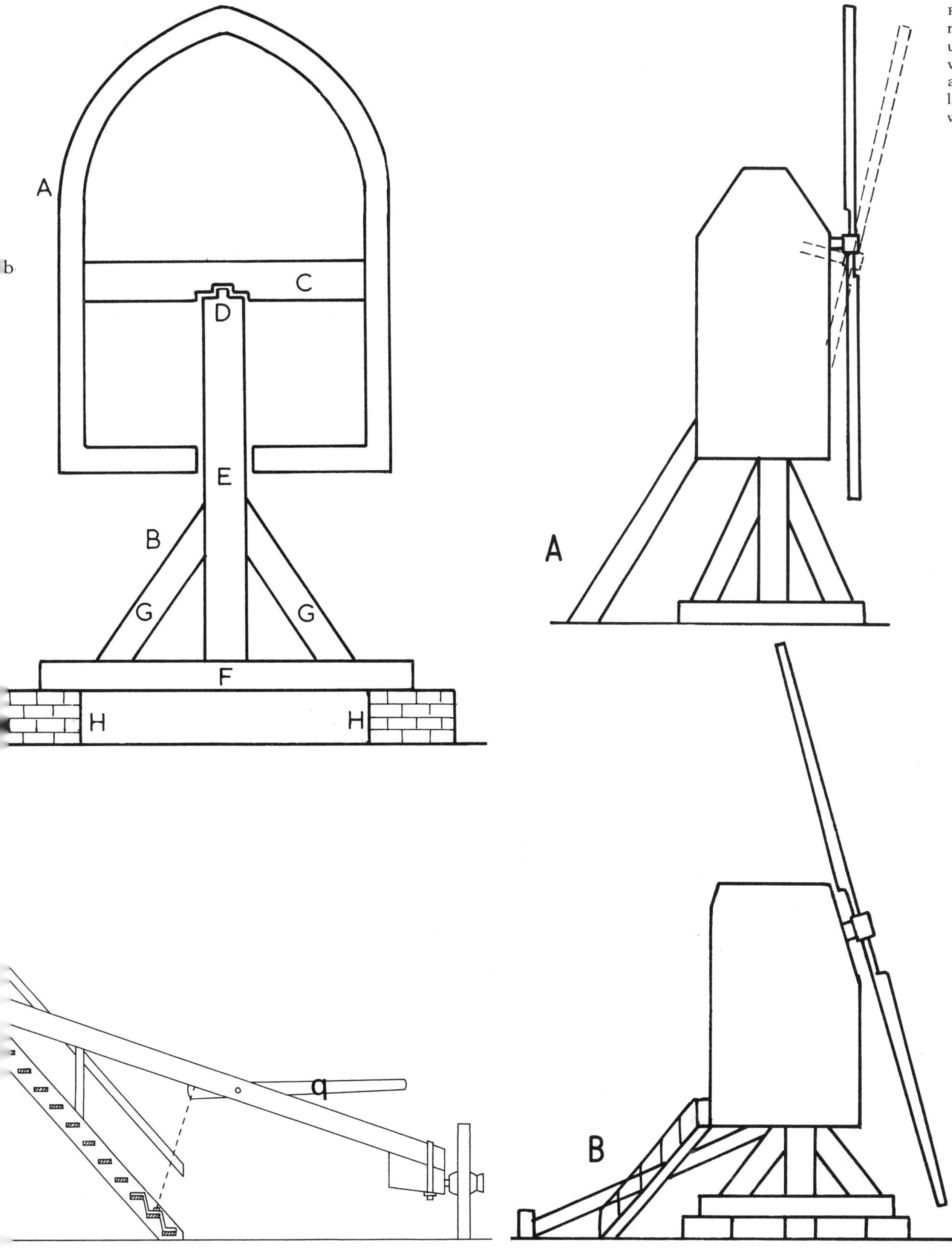

FIGURE 12 Old and new type post mills. A is the early type, clearly unbalanced. Horizontal windshaft will eventually wear its bearings and tip forward. B (below) is the later type obviously well balanced with inclined shafts.

PLATE 43
'MICHEL', COQUELLES, CALAIS
This trestle, with continental double quarter bars, shows clearly the design usual in European and English mills of this type. The kingpost is mortised over the cross trees and held in position with pegs. Although this mill dates from before the Revolution, the post has not yet sagged on to the cross trees, which it clears by about half an inch. The quarter bars are set into the cross trees so that they cannot slip outwards under the thrust of the mill, with deep tenons at right angles into the cross trees, and a 'birdbeak' at the top. The trailing rope works the brake lever high up in the mill so that the miller can stop and start the mill while attending to the sails.

PLATE 44
CHILLENDEN MILL, KENT
Ladder and tailpole set on a wheel to take the weight and make for easier winding. The talthur on the tailpole, which would be hooked to the ladder to lift it clear of the ground when the mill is winded, is the wrong way round in this photograph. Note the overlapping boarding on the front of the buck, and the deep breast to protect the trestle. Iron girders have been built in to stiffen the trestle.

Attached to the bottom of the buck is the tailpole. In early mills this big beam rested at its bottom end on a wooden support or trestle; at Chillenden it is attached to a wheel which takes the weight. Attached to the tailpole is a lever and chain, a *talthur*, with which the ladder is lifted clear of the ground. Then, by leaning on the tailpole and shoving, the body of the mill can be turned to face the wind. Occasionally a horse would be harnessed to the pole to wind the mill.

As post mills were built bigger and heavier, it was a struggle to wind them by man power, and the tailpole and/or the ladder were attached to a wheeled bogie or *fantail carriage* running on a track and carrying a ground level fantail (p. 42).

Post mills are faced with weatherboarding. The back of the mill is covered first, then the sides, with boards projecting slightly beyond the back board to weatherproof the corners. The front of the mill is likewise clad, with the boards projecting beyond the sides, and often projecting downwards in a kind of widow's peak, which keeps some of the weather out of the trestle (plate 4). Painted white or tarred black, there was and is, no finer sight in the English countryside than a post mill – ladder, tailpole and the ground level fantail visually balancing the sails, the overlapping boards casting dark shadow lines, giving perspective and solidity to the shape of the body.

In Chillenden mill, and other open post mills, the trestle is visible, but it was usual to build round it to protect the trestle from the weather and to make storage space. Such a structure, the *roundhouse*, provides a lot of extra floor space for dry storage, very necessary in a post mill, which lacks the several different floors or stages of the tower or smock mill (plate 45). The roundhouse is usually built of brick, with a tiled or boarded roof which runs up under the buck. Some post mills in the north had roundhouses which ran right up to the base of the buck. There was a curb on the top of the roundhouse wall on which rollers on the base of the buck rested, so that it could turn freely. Much of the weight of the mill in this design was taken on the roundhouse wall, relieving the strain on the post.

These post mills, sometimes described as *turret* mills, are really transitional between post and tower mills. It was a logical step to increase the size of the roundhouse to a tower and enclose all the machinery in it, and to reduce the size of the

PLATE 46
IRBY MILL, CHESHIRE
This post mill in the Wirral peninsular was demolished about 1898, but the photograph shows the way the roundhouse wall ran up under the buck. Even two layers of weatherboarding did not prevent the decay of this particular mill.

PLATE 45 (opposite page)
CROSS-IN-HAND, SUSSEX
When this photograph was taken, the mill was in working order, but a few days later, while it was turning, a sail came off, jammed behind the other sails, and caused a lot of damage. The mill is to be repaired, although at the time of writing the fantail is also smashed and the mill is a sorry sight.

PLATE 47
SAXTED GREEN MILL, SUFFOLK
The mill was there in 1706, rebuilt in 1854, repaired and preserved, and is now in the care of the Department of the Environment. Some of the shutters are out, to reduce pressure on the sails, but as the remaining shutters are closed it was clearly turning when this photograph was taken. The fantail on the long ladder is a fine example of its type.

buck until it was a cap containing only brake wheel and windshaft.

Bourn mill has the typical medieval flat-sided roof which did not allow much room inside for the brake wheel (plate 41). The rounded shape of the roof of all the other English post mills is dictated by the shape and size of the brake wheel. Some mills, notably the one at Saxted Green, have seemingly exaggerated and disproportionately tall roundhouses. This is because the mill has been raised – to clear trees or buildings and give it a free wind – by building up the piles under the trestles. Saxted mill roundhouse was raised to three times its original height in this way.

From a constructional point of view, post mills were much easier to build than smock mills, as the rectangular framework could be built up *in situ* on the basis of the crown tree, with mostly square joints and without the intricacy of a curb.

PLATE 48
SOUTH NORMANTON
NOTTINGHAMSHIRE
This sad photograph of a post mill, long since demolished, shows the immensely strong basic construction of the type. The post can be seen going right up through the mill to the crown tree, behind the ladder which gave access to the brake wheel and wallower. The stones are still in position directly below the brake wheel and, although all the weight of the machinery appears to be at the front of the mill, it has not sagged forwards. The trestle appears to be almost indestructible!

4

The removal of windmills

A novel sight was observed at Chelmsford on Tuesday se'nnight, in the removal of the body of a windmill and its contents, entire, from the place where it had been erected many years, upon a scite near the brewery of Messrs Woodcock, Hodges, and Wells. The mill was first divested of its sails and cap, the body raised from the foundation by a lofty triangle, pullies, and blocks, and when sufficiently elevated, a strong four wheeled timber carriage drawn under, upon which it was let down; the carriage and its lofty burden were then gradually removed, the wheels being properly blocked to prevent too sudden a pressure down a short declivity into the road, whence it was conveyed without any accident in the most majestic manner, to the astonishment and admiration of a vast number of spectators, and safely deposited by the same means upon another foundation on a scite prepared for its reception.

Kentish Gazette (21 May 1824)

There are many records of windmills being moved, although it would seem to have been a monumental job. Mills were sometimes moved to better sites to find more or freer winds, or were sold to new owners who may have found it cheaper to pay for their removal than to build new ones. Post mills were the easiest to move. The whole thing, trestle and all, could be transported on a trolley. Usually, however, the buck was jacked up, the quarter bars and cross trees removed, a hole dug and the post dropped into it and taken away; the buck was then lowered on to a trolley to be set up elsewhere, on the same or a new trestle, by reversing the process. Smock mills could also be placed bodily on trolleys or rollers. Brick tower mills obviously could not be moved in one piece; the sails, cap and machinery would be dismantled and transferred to a new tower built on another site.

Some mills were moved with machinery, sails and stones intact, and set up ready for work on the new site – 'instant' windmills, in fact; but it was usual for all movable gear to be dismantled and moved separately to reduce weight. It would take up to forty horses or even oxen to pull a windmill on a trolley to another site, and there are records of all kinds of mishaps. Getting stuck in the mud or on difficult corners occurred frequently, while some mills came crashing off their trolleys on slopes. One can imagine the men yelling and scattering to get out of the way when collapse became inevitable, and the comments of the interested and probably critical groups of onlookers who would inevitably have arrived from somewhere.

On 31 August 1877, the Cross Street mill in Worthing, Sussex, was to be moved out of the town. The mill was owned by Mr Isted, who left us no exact record of why he wished to move it. There were already two mills at Navarino, quite near the intended site at Seamills, whose name indicates that a mill or mills had previously existed there. Possibly Mr Isted was not getting enough business in the middle of the town – there were six, possibly seven mills in Worthing – and felt that by moving out into open country he would get more work.

A Mr Edward Collins dismantled the mill, and moved the machinery, millstones, windshaft and sweeps on ahead to the new site. Once the mill had been jacked up and the trestle and post removed, they too would have been taken to the site to be re-erected if they were sound, or new timbers found to replace them. Mr Isted hired forty horses from a Mr Poland to move his mill – even at a time when horse power was almost the only power available for lugging heavy objects about, to assemble such a team with its harness and tackle must have been quite an undertaking. Imagine harnessing up forty excited horses to be moved off in unison! Pulling windmills around the countryside was not an everyday job, and the sight of a large post mill, even without its sails, pursuing them down the road, could have been too much for the nerves of the steadiest carthorses, so they probably had to be blinkered.

The procession started off from Cross Street, not far from the present site of Worthing Central Station, and passed successfully down Teville Road, Chapel Road and North Street to the High Street, only knocking down one lamp post on the way. So far so good, but when the horses tried to turn their awkward load into Lyndhurst Road, there was trouble. Half the horses went straight on down the High Street, and the mill stuck firmly on the corner. After a lot of pushing and shoving, Mr Poland gave up. One wonders if there were

PLATE 49
MOVING A WINDMILL
This picture of the Cross Street mill being moved out of Worthing, Sussex, was drawn by George Truefitt in 1877.

angry words at this point between Mr Isted and Mr Collins on the one hand, and Mr Poland on the other, for the latter unhitched his horses, packed up and went home. So did the spectators, for the next day promised to provide a lot more fun. Mr Isted had to get the mill out of the middle of Worthing somehow. Imagine the traffic hold-up such a thing would cause today!

Mr Holloway at Shoreham had recently invested in the very latest in steam traction engines for his contractor's business, and Mr Isted was able to arrange for him to come over the next day and complete the job. By this time the news had got right round Worthing, and George Truefitt, a retired architect who drew the picture (plate 49), was on the spot in the morning to record the splendid sight: Mr Holloway himself in a stove-pipe hat driving a traction engine, and the buck of the mill, with its attendant crowd, moving away to its new site. The records clearly state that two traction engines were hired, but there is only one in the drawing, and George Truefitt was always accurate (note the weather vane on the top of the buck). Probably one engine moved the mill without trouble and the other was not needed. It is a little ironic that it took steam power to extricate this windmill from its predicament. The mill continued to work until 1892 and was demolished in 1903; not so very long after all that trouble was taken to re-site it.

Windmills were moved for other reasons too. Villages were steadily expanding, more houses being built and roads widened. A turning windmill would frighten horses, and could endanger anyone who got in the way of the sails. The traffic to and from the mill could also be a nuisance.

The following account describes very well the kind of situations which arose and how they were resolved; £40 would seem to have been a generous sum for the Huntingdonshire miller. It was presumably paid for the removal of the mill to another site and for loss of business while this was carried out.

An indictment was tried at Huntingdon which excited no small degree of pleasantry in the country. A clergyman indicted a miller for working his mill so near the common highway as to endanger life. The clergyman is a man of considerable property and consequence in the county. He was obliged daily to pass by this road on horseback, and had been several times thrown by his horse taking fright at the sails of the mill. The mill was an old one which had formerly stood on a common, but the latter had been enclosed by the Commissioners under the Inclosure Act, who had directed the new highway to be so placed that it passed close under the fly of the mill. Mr Justice Grose, addressing the jury, said the mill as it stood now was unquestionably a nuisance, and the miller must be found guilty. It was, however, no fault of his. The Commissioners who directed the road to be set out were most to blame, and he regretted they had not been made parties to the indictment. Neither was the prosecutor to blame in preferring the indictment; he could go by no other road, and his life as well as those of his fellow-subjects was endangered. Under such circumstances he felt wholly at a loss how to act. The miller ought not to be punished; yet the safety of the prosecutor and the public must be consulted. He thought the best way of deciding would be to direct the *prosecutor to pay the miller £40* and the miller to abate the nuisance, with leave to erect his mill on some convenient spot adjoining. This was accordingly made the decision of the Court.

Annual Register (18 July 1807)

PLATE 50
'JILL', CLAYTON POST MILL
SUSSEX
She was built at Dyke Road in Brighton in 1821 and moved up to the top of the downs by a team of oxen in 1866. She has a pitch-pine main post.

5

Windmills in Denmark

The National Museums of Denmark formed a Mill Preservation Board in 1953 (see Appendix 2). Its first concern was to save the 10 per cent or less of watermills which were still not totally ruined. Although there had once been several thousand windmills and many were still in reasonable condition, very little measuring had been done or basic records kept. The Mill Preservation Board, led by Anders Jesperson, a world authority on mills, is extremely active. At the time of writing there are seventy Danish windmills scheduled for preservation, the main effort at the moment being to save the worst from complete destruction – to carry out a holding operation until they can, one by one, be more completely repaired. Dyrehavens Mølle, the subject of figs 13 and 14, ceased commercial work in April 1973 for economic reasons. It is possible that in Denmark as elsewhere energy and fuel problems might help to save windmills capable of being converted to produce power. It's an ill wind that blows nobody any good!

Watermills dominated the scene in north-western Denmark, while windmills were more

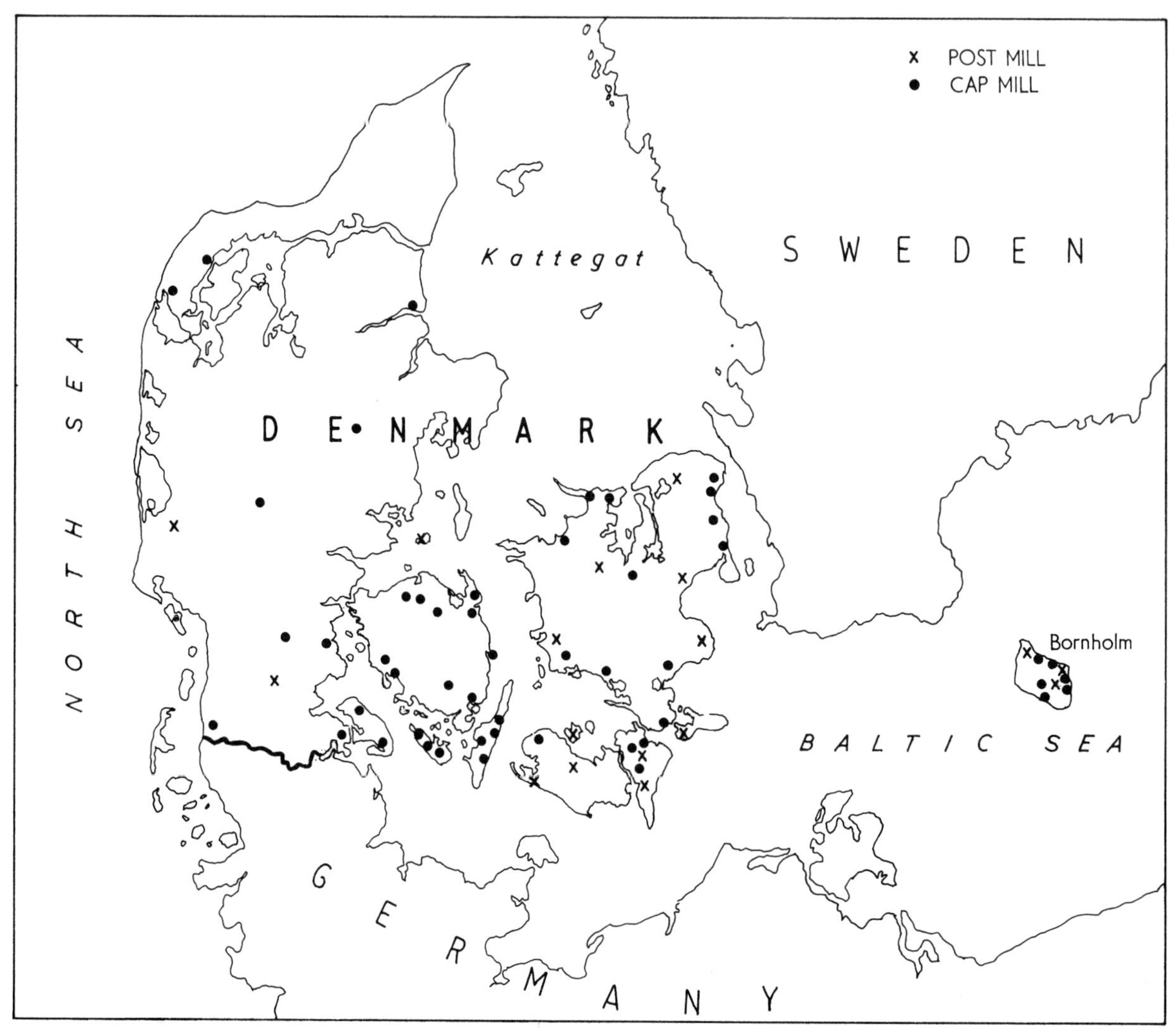

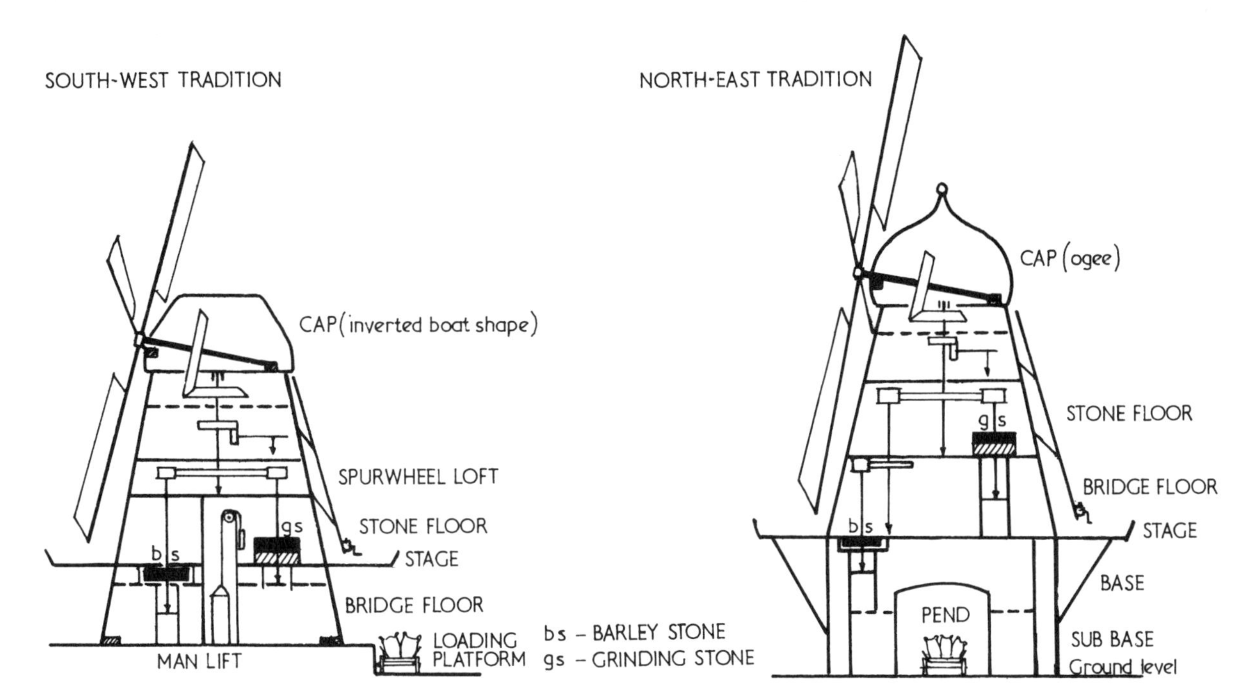

FIGURE 13 Types of Danish mills.

common in the south and east. In 1862 Mølle-privillegium, which had much in common with the English Mill Soke, was abolished, and hundreds of windmills were promptly built to compete on a free market. This seems to have been a case of everyone getting in on the act; the windmills were not always built by millwrights, but by local builders, and the standard of construction deteriorated. Milling followed the same pattern in Denmark as elsewhere. Power-driven roller mills in the ports dealt with flour, and grinding for animal feedstuffs moved to individual farms. Small millstones and other machinery would be powered by a pair of sweeps on the barn roof, or by small sails built by the village smithy. Eventually engines superseded these. Yet, where mills still exist, they are again being used because today Danish farming suffers from a labour shortage.

Seventeen early post mills survive, two re-sited in museums in Jylland, the south and west of mainland Denmark. The rest are in the east of the country, many on the islands. They all have direct drive to the stones from the brake wheel.

This molinological map, which shows the distribution of the different types of windmill, though not their state of preservation, gives a pretty good idea of where to find windmills in Denmark.

The later tower mills are impressive affairs, especially the Scanian or north-eastern type, with its brick-built base and stage, and the mill-proper above it. The basic similarities and differences between this type of mill and those in Britain are clearly illustrated in the diagrams. Most noticeable are the sails, with their automatic shutters pivoting on the stocks. The drive mechanisms differ only in detail from those of British mills built at the same time. The Frisian or south-western type was built without the brick base and had a boat-shaped cap, not the ogee of the north-eastern mill. There were any amount of intermediate stages between the two types. Until 1850 all the big mills were octagonal, made of wood and faced with thatch (like Dutch mills) or shingle. After that date, especially in the south-west, where bricks were readily available, they were bricked right up to the cap.

The fantail is as common as in English mills. A firm of English millwrights is known to have done a lot of work in Denmark, and possibly they introduced the idea to the Danes. When this automatic winding was introduced, as late as the 1920s, the mills were given cast-iron curbs to carry the caps and take the strain of the continual slight movement. But cast iron is extremely brittle, not of the same elasticity as wood, so the curbs broke and the mills were wrecked or abandoned. Many smock mills were lost in this way. Nowadays, when a curb has to be replaced, it is cast from mild steel, which is nothing like so brittle and will stand considerable movement and subsidence in its wooden windmill before it breaks.

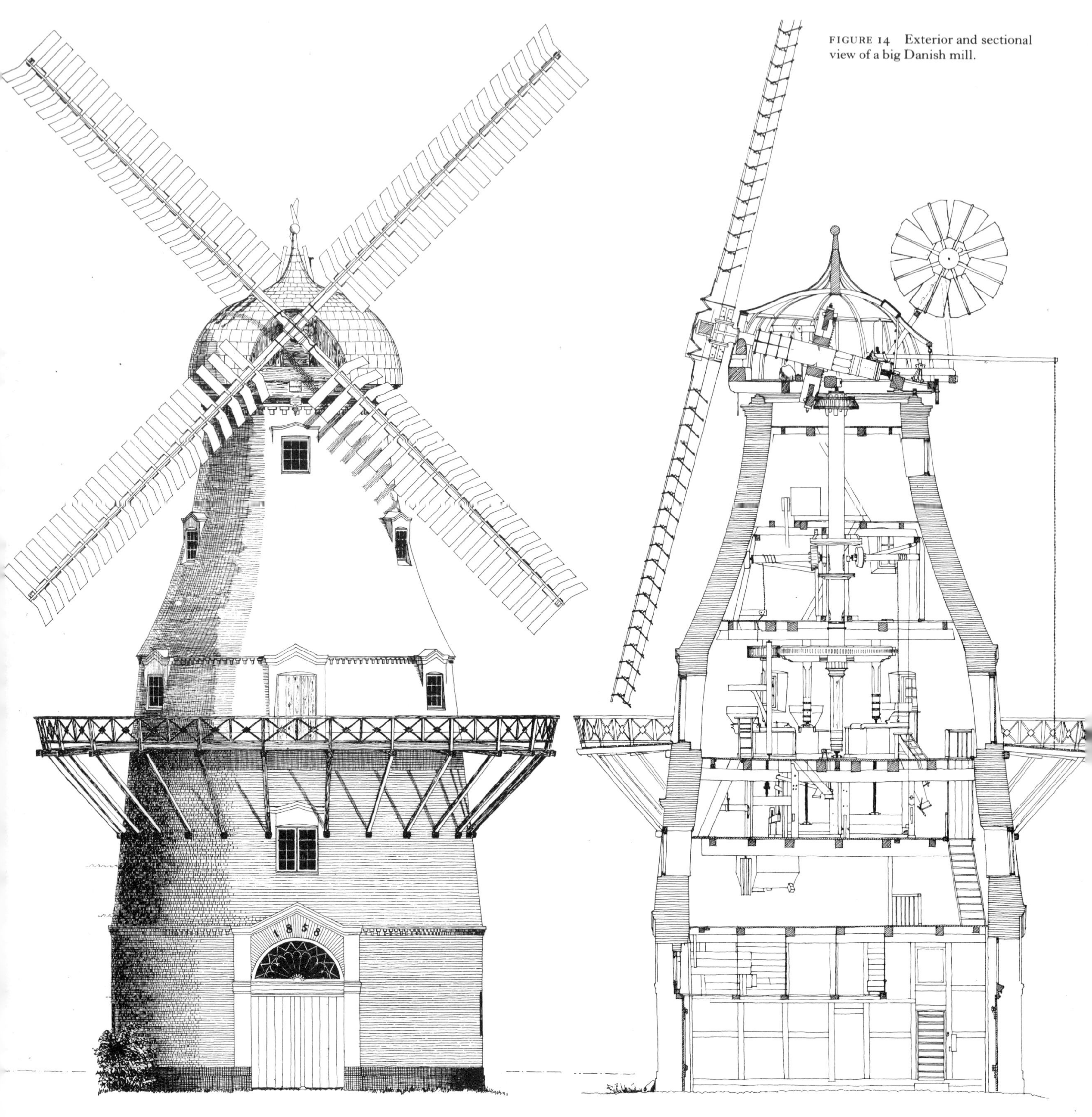

FIGURE 14 Exterior and sectional view of a big Danish mill.

6

Millstones and milling

Behold! A giant am I,
Aloft here in my tower
With my granite jaws I devour
The maize and the wheat and the rye,
And grind them into flour.

I look down over the farms;
In the fields of grain I see
The harvest that is to be;
And I fling to the air my arms,
For I know it is all for me.

I hear the sound of the flails,
Far off from the threshing floors
In barns, with their open doors,
And the wind, with wind in my sails
Louder and louder roars.

On Sundays I take my rest;
Church-going bells begin
Their low, melodious din;
I cross my arms on my breast,
And all is peace within.

Longfellow

Millstones

The earliest type of true millstone was a *quern*. This consisted of two stones, the bottom one being a hollow, thick disc into which the top stone fitted. A hole was bored through the centre of the upper stone by spinning a pointed hardwood stick with a bow string and pouring wet sand round it as an abrasive. This must have taken a lot of time and patience, for the quartzite was very hard. Another smaller hole was bored near the rim of the stone to take a piece of wood as a turning handle, and the grain to be milled was fed in through the centre hole or *eye*. As the top stone was turned, the grain would be crushed and ground into some semblance of flour. Every so often the top stone was lifted off and the meal scraped out.

The earliest windmills turned what were little more than querns, but very soon millstones were developed in the form in which they have remained to the present day.

The best and most expensive millstones came from the Rhine valley, but cheaper substitutes were found. In Britain, millstone grit, sandstone from Derbyshire and the West Midlands, and granite from Dartmoor and Wales were all used for stones. 'Cullen' stone from Cologne, or Köln, remained the best, except for *French Burrs*, which were made from blocks of a special quartz quarried near Paris. This stone did not come in big enough lumps to make millstones in one piece; small blocks were very carefully shaped and matched, then lightly cemented and bound with iron hoops to hold them together.

Millstones average about 4 ft or a little more in diameter, and 8–14in thick. Each stone can weigh up to 1½ tons.

The *bedstone* is set perfectly level with its face a few inches above the floor (plate 53), inside the *vat* or *tun*, an octagonal or round wooden box slightly bigger in diameter than the stones. The stone has a central hole with a special bearing box just big enough to take the *stone spindle* which comes up from the *bridge tree* below to support the *runner stone*. This upper or runner stone has a big hole – the eye – through the centre. Cemented across the lower side of this hole is the *bridge*. In the middle of the bridge is a slight depression or dimple into which fits the very tip of the stone spindle. There is a heavy cast-iron coupling, called a *mace*, on the top of the spindle, slotted so that it fits round the bridge, and when the spindle turns the bridge and runner stones turn with it (fig 8). An overdriven runner stone is turned by a quant coming down from the stone nut. This ends in a Y-shaped crotch which slots over the bridge and into the mace from the top. The stone spindle from below still supports the runner stone and turns with it. Thus the runner stone is suspended just above the bedstone, so close to it that when it turns it will crush and cut up grain between itself and the bedstone, but not so close as to touch it.

To balance the runner stone properly, it must have lead run into cavities on its back, or special balancing weights let into recesses near the rim of the stone (plate 53). To raise or lower the stone, there is an arrangement of screws and levers under the floor, in the next lower stage of the mill. The bridge tree, which has a footstep bearing in it to take the lower end of the stone spindle, rests on

PLATE 51
MILLSTONES AT
RICHBOROUGH CASTLE, KENT

Although lying with a heap of Roman masonry, these millstones are not so old, but the broken one at the top does not even appear to be completely circular and could be quite ancient.

PLATE 52
UNION MILL, CRANBROOK, KENT

The vat containing a pair of millstones. A wooden shovel was used because a metal one would bite into the wooden floor and lift splinters as well as grain. In front of it, the manyheight, a step-shaped wedge, served as a fulcrum for a crowbar to help when moving stones for dressing.

PLATE 53
PAIR OF MILLSTONES AT
HERNE MILL, KENT
Note the slot in the side of the runner stone to take a special Lewis bolt for lifting the stone. The bridge is in position in the runner, and the bearing box is in the bed-stone. The stones are made of carborundum composition.

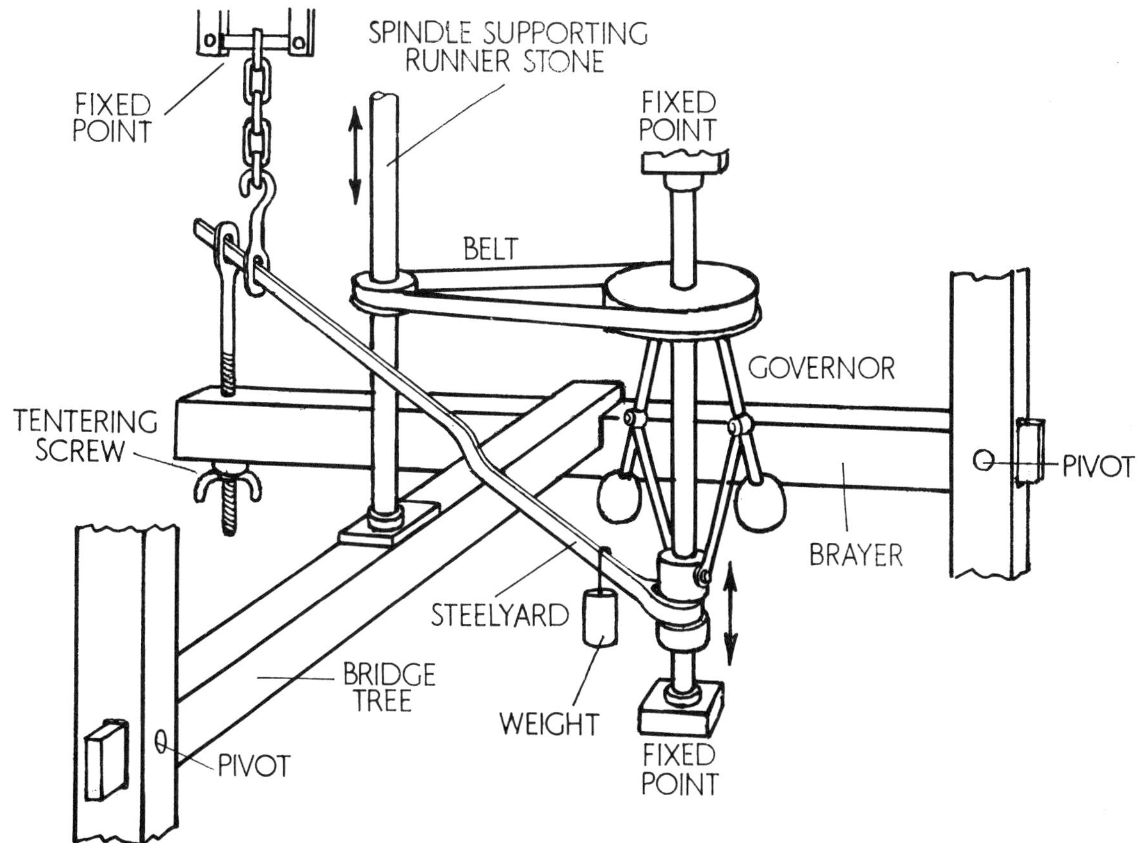

FIGURE 15 Action of governor and tentering screw (diagram not to scale).

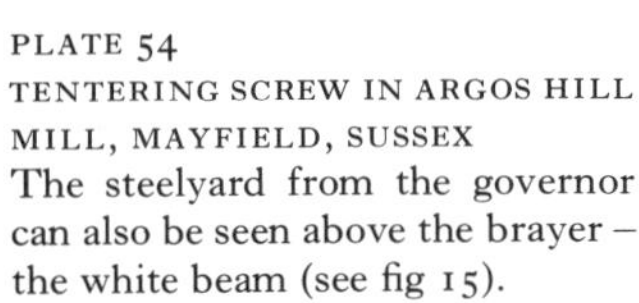

PLATE 54
TENTERING SCREW IN ARGOS HILL MILL, MAYFIELD, SUSSEX
The steelyard from the governor can also be seen above the brayer – the white beam (see fig 15).

PLATE 55 (above right)
GOVERNOR IN STELLING MINNIS MILL, KENT
Note the Y-shaped ends of *two* steelyards. This governor could control two sets of stones at the same time.

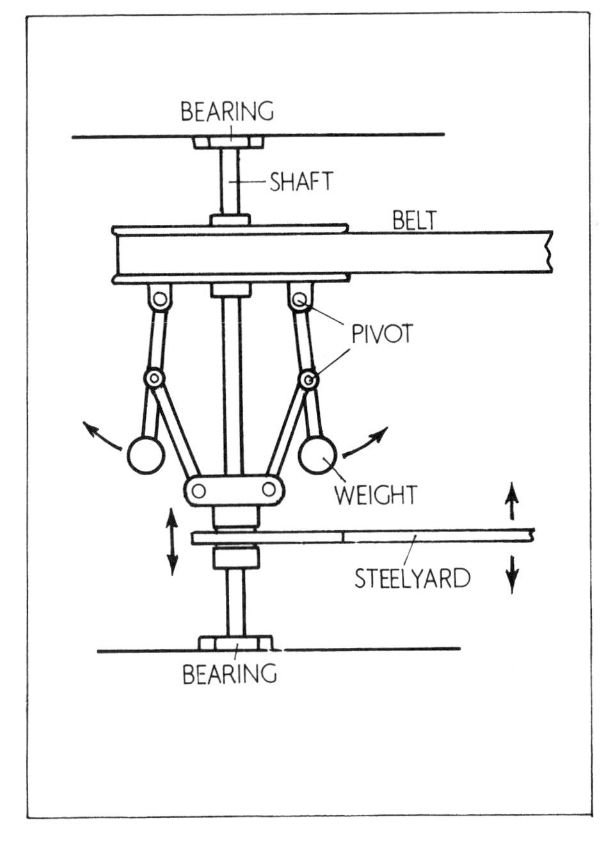

FIGURE 16 Governor.

another beam set at right angles to it, the *brayer*; at one end of this is a *tentering screw*, by which it can be raised or lowered, thus raising or lowering the bridge tree, the spindle and the stone. Some mills have no brayer, only a tentering screw acting directly on the bridge tree. Here also is located the *governor* which works from a belt drive usually from the stone spindle. As this governor goes faster or slower, it automatically raises or lowers the runner stone fractionally by means of a *steelyard* running to the brayer above the tentering screw. The gap between the stones must first be set with the tentering screw. The faster the spindle turns the more the governor lowers it, allowing the runner stone to drop closer to the bedstone and grind more efficiently the extra grain which is passing through the eye. As the mill slows, the governor drops and the spindle rises again to lift the runner stone to its widest setting.

Stone Dressing

This is the recutting of the grooves – known as *furrows* and *lands* – in the working surfaces of the millstones.

Before the runner stone is dressed it has to be lifted off the spindle and bedstone, with slings and blocks and tackles, and laid flat on the floor. Sometimes there was so little room that the stone had to be leaned at an angle which made it even harder to dress, or it was hoisted on edge and turned upside down on top of the bedstone. It would then be pulled up on edge again while the bedstone was attended to. In plate 52 is a *many-height*, a stepped wedge shape used as a fulcrum for a crowbar to help lever the stones out.

The stone dresser has several tools. First, the *wood proof* or *paint staff*, a mahogany straight edge, and a *steel proof* – a flat steel bar kept in a mahogany case – against which the wood proof must be continually checked for accuracy. Two small pieces of batten known as *furrowing strips*, one the width of the furrow, the other the width of the land, are used to mark out the lands and furrows, with the aid of a feather dipped in *raddle*, and to measure their depth. A tin of raddle or *tiver*, a composition of red oxide (some millers used blue dye made from woad); mill *bills* and *thrifts* or *stocks*. Mill bills are double wedge-shaped pieces of steel, specially tempered and sharpened, which fit right through the hole in the top of the thrift or handle and are wedged in place. Lastly, a *bist* or rough cushion to lean on while working across the stones.

The bailiff purchases iron (2½d) and steel (for bills, chisels and picks to bore the stone, 9d) hires a smith (2s) to fashion the picks, and engages the services of three men for three days (3s 9d) in the labour of boring the stones – a labour of no trifling character, as the smith is perpetually employed sharpening the picks.

The Bailiff of Cuxham Manor, Oxford, 1330

PLATE 56
STONE-DRESSING KIT IN STELLING MINNIS MILL, KENT
The millwright is holding the steel proof; the wood proof lies on the millstone, with the furrowing strips, and also a thrift which has a very short and nearly worn-out bill in it.

The wood proof is smeared with raddle and passed across the surface of the stone, the red dye transfers to the stone and clearly shows the raised parts and depressions. The surface is first made perfectly smooth but the runner stone is dished slightly towards the eye, giving a 'swallow' of about a quarter of an inch to prevent uneven feeding of the stones.

Next, if the stones are to be used for grinding flour, the furrows and lands are cut. These are a regular pattern of grooves which allow the grain to move out to the edge of the stones as it is ground. Each furrow, cut about half an inch deep has a deep edge or arris on one side and slopes up to a sharp shallow arris on the other. It is this arris which does a lot of the cutting. Lastly, a series of fine parallel grooves – drills, feathering or stitching – about twelve to the inch, are cut along the lands; it is these which do most of the grinding work. This operation, known as *cracking*, has to be done frequently if the mill is working most of the time. In the first third or *skirt* of the stone, these are cut deeply; in the middle third or *breast*, a little shallower, to die out in the *eye*, the centre third (plate 53). A really good dresser can achieve as many as sixteen grooves. The furrows are cut in the same way on each pair of stones and, when the runner stone is turned over above the bedstone and is started, the furrows cross each other like scissor blades; it is this action which grinds the grain. Some types of millstone are so rough and hard that cracking is unnecessary, the rough surface being sufficient in itself. These are known as *open stones*.

A pair of stones will grind between 300 and 400 tons of grain before needing to be dressed. The harder and better the stone, the longer it will remain sharp; French Burrs will grind considerably more than this tonnage. It takes up to 100 hours to dress a pair of stones properly, although a full-time stone dresser could do it much more quickly than this. So it is clear that it took a high percentage of total working time for the miller to keep his stones in order and that at least two pairs would be needed so that they could be changed over to keep the mill grinding.

The chipping of the stone is done with a bill held in a thrift. To do this work the dresser sits on the edge of the stone, resting his body and arms on sacks or bists. The bills must be sharpened constantly on the miller's grindstone. When they get too worn down they must be taken to a blacksmith who hammers them out again and tempers them ready for re-use – a job some smiths do much better than others. Bad tempering results in the bill chipping as it is used. In any case, little bits of metal occasionally fly off and embed themselves in the dresser's hands. If an unknown dresser arrived at a mill looking for a job, he might have been asked to 'show his metal' to prove he had been at the work long enough to be skilled at it.

PLATE 57
The millwright holds the thrift and also the tracer which is used to check that the spindle is running true.

After dressing, the stone is tested for level with a spirit level. The *tracer* (a special piece of wood with a quill stuck through the end dipped in raddle) is then slipped over the exposed spindle to check that the spindle is running true. Then the sails are allowed to turn slowly and as the spindle turns the quill scratches on the bedstone. It should leave a perfectly centred circular mark, and make an even scratching noise. Adjustments are made to the footsteps bearing on the bridge tree until this is achieved.

FIGURE 17 Arris.

Drills Arris Arris

Grain Feed Mechanism

When the runner stone is back in position, the wooden vat is put back into place. On this is built a horse which carries the *grain hopper* (plates 52).

PLATE 58
Close-up of the thrift and mill bill.

There is a sliding door in the hopper which is set by hand to regulate the flow of grain into the *feed shoe*, a sloping wooden trough, free to move from side to side, which feeds the grain down into the eye. Sticking up from the top of the mace, or in some cases the bridge, is a four-sided rod, the *damsel* – the noisiest thing in the mill! The shoe has a *rap* which bears against the damsel; as the damsel turns, it pushes the rap from side to side, shaking the shoe so that grain dribbles down into the eye. On overdriven stones the four-sided quant itself acts as a damsel. The faster the stones turn, the more the shoe shakes and the more grain is delivered through the eye. A *warning bell*, which rings continuously if the grain in the hopper gets low, is usually incorporated; it is set so that the miller has a few minutes before the hopper runs dry to get up another sack and refill the hopper. This bell is attached to a strap long enough to loop down into the hopper with the grain weighting it. When the grain gets too low and no longer holds down the strap, the weight of the bell pulls up the strap and the bell drops down to rest against any moving part, ringing loudly.

PLATE 59
UNION MILL, CRANBROOK, KENT
Looking down into the eye of the runner stone. The block, or rap, on the shoe bears against the four-sided driving quant with its Y-shaped crotch slotting down into the mace, across the bridge. The shoe is held under tension by a leather thong attached to a metal spring or wooden stick – the rabbet. The purpose of the stick wedged into a clamp and passing into the eye is to keep the eye clear of grain which might build up there.

Milling

. . . All the science and tact of a miller is directed so to regulate the machinery of his mill that the meal produced shall be of the most valuable description that the operation of grinding will permit under the most advantageous circumstances. With stone-grinding the miller's ear is constantly directed to the hum made by the running stone, its exact parallelism to the bed-stone, indicated by a particular whirring sound, being a matter of the first consequence. At the same time his hand is constantly placed under the meal-spout to ascertain by actual contact the character and quality of the meal produced. The thumb by a particular movement spreads the sample over the fingers, and the thumb is the gauge of the value of the produce. By this incessant action of the thumb is produced the peculiarity of form, said to resemble exactly the shape of the head of the fish so often found in the mill stream.

William Yarrell, *History of British Fishes* (1835)

As the grain is ground between the millstones the crushed material moves down the furrows of the millstones and out towards their edges to fall into the wooden vat surrounding the stones. Attached to the side of the runner stone there may be a wooden sweeper arm or *tag*, which sweeps the meal along the vat into the meal hole, whence it departs down a chute into the meal bins below. The meal, warm from the process of milling, and filling the whole mill with the rich pungent smell of freshly broken grain, sometimes has to be turned with shovels to let the air in and cool it down, but this is bad milling.

Clive Wootton, the miller of Herne, reckons that it is impossible to teach anyone to be a good miller unless he has a natural flair for the job. It takes years of experience to be able to judge almost at a glance and a touch the qualities of the wheat to be ground, whether it is wet or dry, and what sort of flour it is likely to produce. The secret of milling is to keep the stones running to capacity all the time, but never to choke them or to run them half empty. A miller can walk into his mill and tell by the smell whether or not the stones are running too 'hot'. Flour coming too warm from the stones will never run sweetly, 'floating out of the scoop at a gentle shake'. Flour ground too hot seems to cling together and drops in dollops from the scoop.

During World War I, millers were required to produce a percentage of white flour which was difficult for them to achieve either practically or economically. It was this fact which led to the rapid decline of the windmill for grinding anything but animal feedstuffs. Today the power-driven roller mill does the work, but does not produce the fine quality flour made by a good pair of millstones and a miller who knew his job. One miller was heard to remark that the roller mill does not grind the flour out of the wheat, 'it frightens it out'.

If wheat is merely crushed, the brown skin flakes off as *bran*; but in the windmill the wheat berry is ground down to a fine wheatmeal which can be used to make rich brown bread with all the nourishing parts of the wheat still in it. To dress the wheatmeal – that is, to turn the wholemeal flour into white flour – it must be passed through a *flour dresser* or *bolter*; this takes its drive by means of belt and pulley from the main driving mechanism.

One type of flour dresser has a silk sleeve woven in three sections: fine, coarse and very coarse. It is stretched over a wooden frame cylinder mounted on a spindle and inclined so that meal fed into one end passes right down the drum made by the silk sleeve. The cylinder or reel spins fast and is enclosed in a framework containing longitudinal wooden bars or beaters. As the flour in the spinning drum presses up against the silk it makes it belly out against the beaters and the flour is rubbed through the cloth. The finest flour comes out immediately and falls into the first of four hoppers. Second-grade coarser flour passes through into the second hopper. Into the third hopper passes the *supers*, used for animal fodder, with next to no flour content, and out of the end of the drum trickles the coarse bran into the fourth hopper.

This type of dresser produces the finest grades of flour, but became useless with the invention of the combine harvester. Combined grain, as opposed to threshed grain, is so full of little bits of straw and other rubbish that it really needs to be cleaned before it is ground. This would necessitate more expensive machinery – for which, in any case, there is no room in a windmill. The *silk-sleeve dresser* at Herne, which produced beautiful flour, was hardly used and eventually thrown out because it just did not get the quality of grain it could deal with.

Another type of dresser works on exactly the same principle, but has a leather-bound sleeve of woollen bolting cloth and while the flour passes out through it there is no grading – all the bran and *middlings* trickle out together at the lower end.

A common type of flour dresser is the *wire machine* in which the drum cannot rotate and

PLATE 60

WHOLEMEAL FLOUR

WOOTTON BROS.

Millers and Corn Merchants

HERNE MILL, HERNE BAY

TEL.: HERNE BAY 1537

And at 19 EAST STREET

3 LB.
GROSS

HOBBS, MAIDSTONE

PLATE 61
NONINGTON MILL, KENT
The flour dresser – a wire machine – slopes down at the back of the photograph. In front is the crown wheel and pinion drive to a horizontal shaft which carries pulley wheels for belt drives to other subsidiary machinery. The crown wheel is at the bottom of a vertical shaft taking its drive from the great spurwheel on the floor above.

contains sections of wire mesh in progressively coarser gauges. Down the centre of the drum passes a spindle carrying brushes which, as the spindle rotates, force the meal, fed in at the top end, against the mesh. This action sieves it successively into first, second and then third grade flour, then into middlings or *sharps*, then *pollards*, and then into bran.

Now then, to begin with the bread, a pound of good wheat makes a pound of good bread; for though the offal be taken out, the water is put in; and indeed, the fact is, that a pound of wheat will make a pound of bread, leaving the offal of the wheat to feed pigs, or other animals, and to produce other human food in this way.

William Cobbett, *Rural Rides* (30 August 1826)

7

Dutch windmills

No other country has so many windmills as Holland. Everywhere across the flat lands, under the wide bright skies there are hundreds of huge, solid, strong windmills, like the Dutch themselves, dominating the landscape in which they live. Even in stillness they are impressive, but turning in a stiff breeze from the cold North Sea these big mills give an impression of power unequalled by any other man-made machine.

Of old, the Dutch lived in a country constantly menaced by water – by the sea and by the rivers of Europe flooding the low-lying land. They used the wind, the other great element which swept round them, to fight the water. By converting their grinding mills (first recorded early in the thirteenth century) to pumping mills (first recorded in 1414), they began the work – which is still going on, although no longer by windpower – of reclaiming huge areas of land from the sea. The windmill continued to be used for drainage of reclaimed land up to the present century, when oil and steam engines gradually replaced the mills.

The Dutch realised that, as each mill became derelict, a vital part of their heritage was disappearing and completely altering the look of their countryside. One irreplaceable complex of fifty was dismantled in the 1930s to make way for modern pumps. The new polders, made with modern machinery not windmills, have a strangely bare look compared with the old windmill areas of Holland. Of the 9,000 windmills which existed in the nineteenth century, about 1,000 remain today, and the work of restoration and preservation continues.

To drain a large area, the Dutch first dug a dyke and a ring canal round it, and then at intervals installed windmills to lift water from the lake into the ring canal. The height of the lift is governed by the radius of the scoop wheel, usually about 5 ft (diameter 10 ft). The mills would work away until they had lowered the level of the lake by 5 ft. Then another series of mills were installed at that level to lift water from 5 ft lower down to a series of catchment areas; from there the original mills would pump it into the ring canals. This was repeated until the deepest pools were drained. The water in the ring canal, at a comparatively high level, could then be drained away by gravity, or if necessary by further pumping, into the river systems, and eventually into the sea. In this way *molengang* sets of windmills were built up to drain the polders into the canals and rivers. At Kinderdijk, near Alblasserdam, there are sixteen polder mills all together. They do not work all the time, although once a year they are all set going – a truly wonderful sight. One mill is always kept working for the benefit of tourists.

The earliest pumping mills were hollow post mills. They were developed from the ordinary type of trestle post mill used for grinding throughout northern Europe, which varied little in principle and only in regional detail. However, to transfer the drive of a post mill to a scoop wheel was obviously difficult because the whole body of the mill turned on its post while the scoop wheel must remain fixed in its basin. To overcome this problem, the Dutch balanced the mill on a hollow post and carried the drive down through this by means of a separate shaft. The wip mill has a buck on a tall post, proportionally smaller than that of a trestle post mill, and round the post is a thatched house enclosing the crown wheel at the bottom of the vertical shaft and the vertical pit wheel geared to it. The rest of the roundhouse is taken up by the miller's living quarters. It always has two doors so that one can get in and out whichever way the buck and sweeps are turned. But there is not much room in there, so if the miller has a big family he needs a cottage as well!

Next evolved the very much bigger polder mill. This is an octagonal smock mill with thatched sides (thatching reed was, of course, available everywhere in the marshes) and a moveable cap (plate 6). A polder mill has a brick or wooden base and inside this the scoop wheel. The thatching on a wooden frame meant that the mill was much lighter than if it were made entirely of brick; on the soft Dutch subsoil this was an important factor. The concave sides of the polder mill look very beautiful, but there is a practical

FIGURE 18 Draining polders.

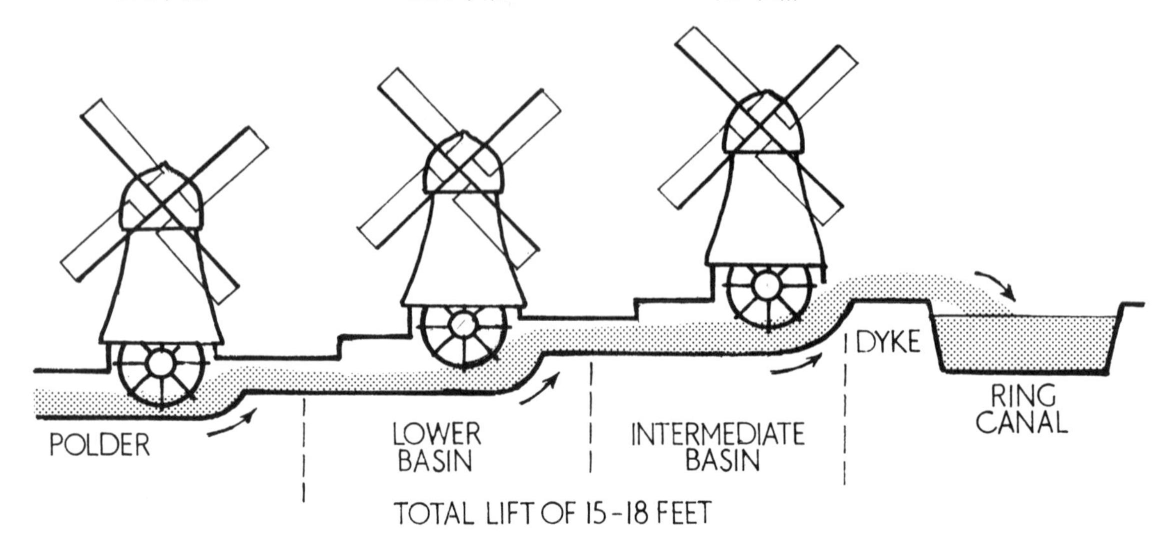

FIGURE 19 Wip mill.

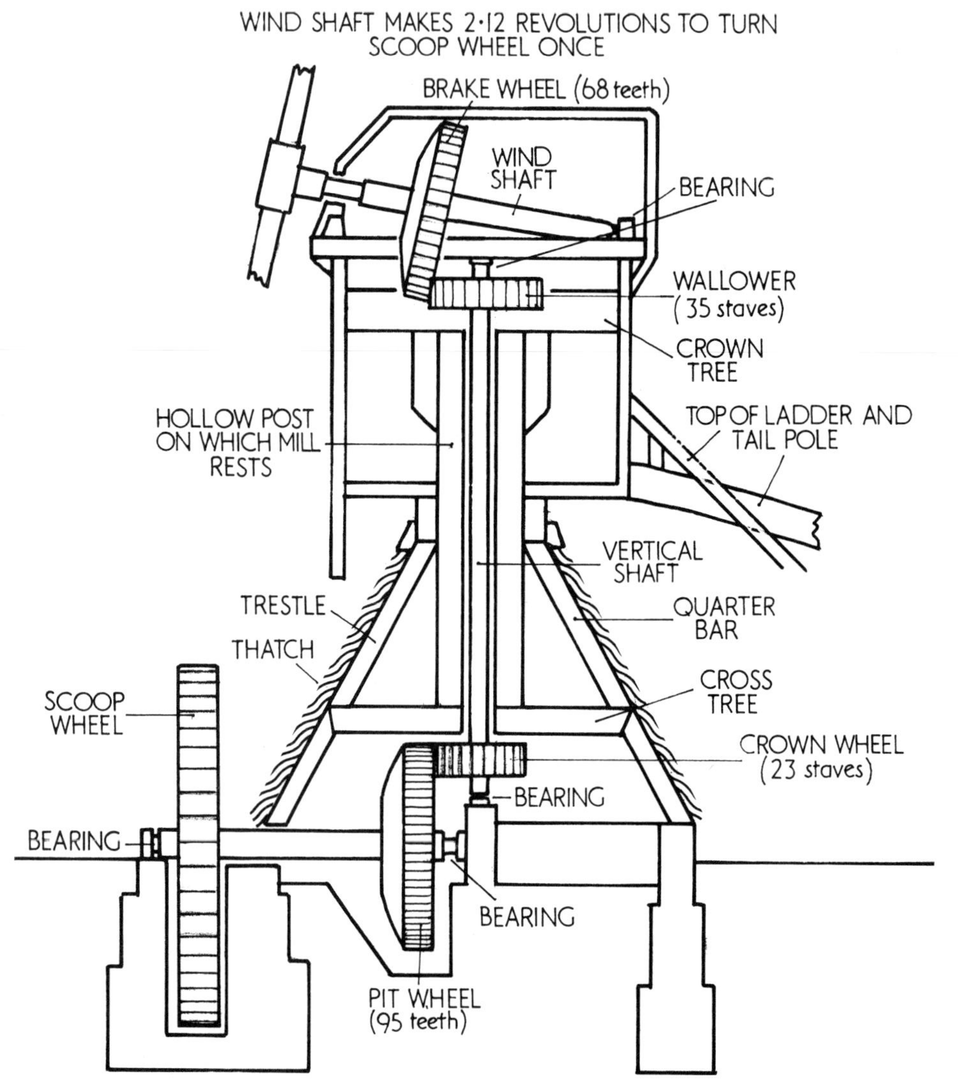

FIGURE 20 (opposite page, above) plans of a polder mill.

PLATE 62 (opposite page, below) POLDER MILLS AT KINDERDYJK HOLLAND

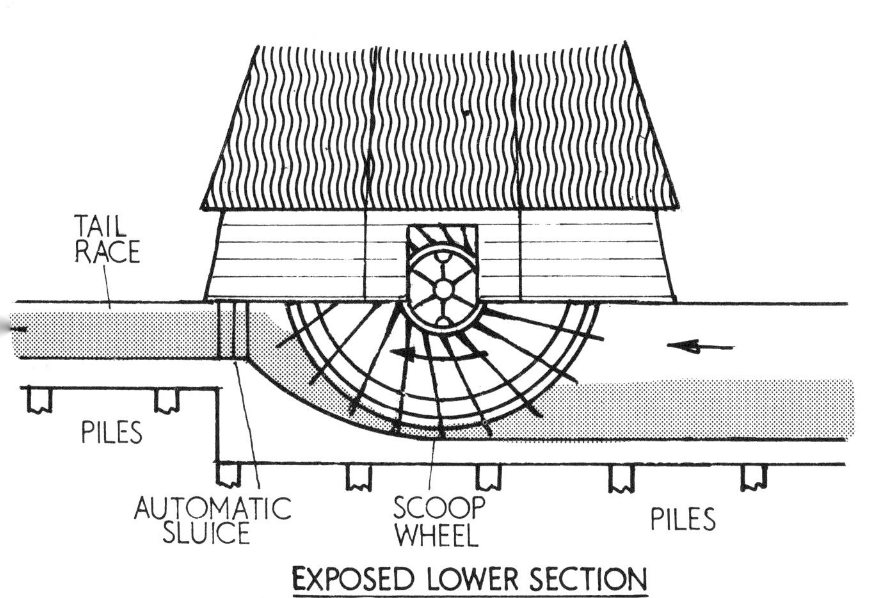
TAIL RACE
PILES
AUTOMATIC SLUICE
SCOOP WHEEL
PILES
EXPOSED LOWER SECTION

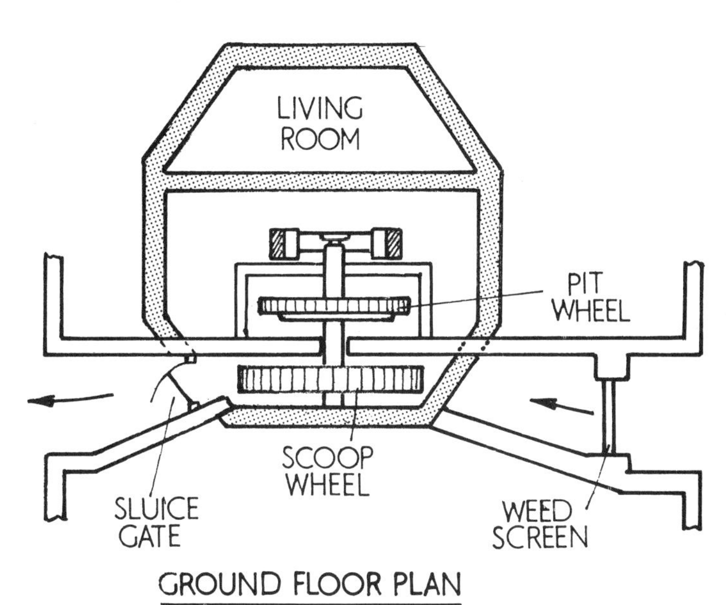
LIVING ROOM
PIT WHEEL
SCOOP WHEEL
SLUICE GATE
WEED SCREEN
GROUND FLOOR PLAN

PLATE 63
DUTCH OPEN POST MILL
This mill was recently reconstructed on the ramparts of the old city of Heusden, in Holland.

reason for their incurving shape: the wind, pushed ahead and sideways by the descending sail, meets less resistance than from a straight-sided mill.

Because a polder mill contains very little machinery in relation to its size, half the ground floor can be used as a living room and the first floor is the bedroom.

Dutch mills carry a decorated board on the front of the cap, called a *beard* (plates 6, 20). This bears the date of the building of the mill and is usually beautifully carved and painted. Despite the gay colours with which the Dutch paint their mills, common sails of dark canvas are usual and show up in black silhouette for miles when set; this gives Dutch mills a much more menacing look than those in England with their white shuttered sails.

Apart from the wip mill and the polder mill, which are specifically drainage mills, several other types are common. To grind corn, the ordinary post mill of the Low Countries – with either an open trestle or a roundhouse, and a tall buck – was used from the earliest days. In Holland it frequently had a thatched cap.

Brick tower grinding mills, again with thatched caps, also survive. Most impressive of all are the colossal tower mills, each encircled by a working stage half-way or even higher up the tower. Because Holland is so flat, small man-made mounds were raised to carry windmills; even then – especially in towns where other tall buildings took too much wind away from the mill – the Dutch had to build high. In these huge mills, the bottom of the tower made a splendid house for the miller; above the circular stage, the brick tower would be continued to cap level, or a smock mill – outwardly like the polder mill – with concave thatched sides, would be built on top of the brick tower.

There were also various types of grinding mills for chalk, dye, cocoa, tan, spices and cement; and mills which pressed oil from mustard and other seeds, leaving a residue of hard cake to be used for animal feed.

An *edge mill* was used to grind chalk, etc, and also seeds for oil production before they were put through the stamping machinery (see below). The mill consisted of two runner stones set on edge on a horizontal spindle which passed through the main vertical spindle from the stone nut. These two stones, one set nearer to the central spindle than the other, rotated and rolled upon a solid pan. The material on the pan was guided into the stone by wooden guides and, when it was sufficiently pulverised, the gate could be opened through which it could be brushed into a receptacle below.

In the *oil mill* a vertical cam wheel on a heavy horizontal shaft was driven by the crown wheel at the bottom of the main vertical shaft coming down from the wallower. This cam shaft rotated the cams thus causing a series of rams and stamps to rise and fall upon the meal (from the seeds previously ground in the edge mill), which had been heated and put into a special bag. It would first be pressed by means of wedges beaten by the rams, then taken out, broken up into pieces and put into pots beneath the stamps.

Sawmills were also common in Holland, and can still be found. The smaller ones used the wind to turn circular saws, but the larger mills could cope with the heaviest sawing jobs.

In the sawmill, a crank wheel on a horizontal crankshaft takes power directly from the crown wheel. Long connecting rods from the crankshaft pass down through the mill to the sawing floor, where they are fixed to the top of heavy saw

PLATE 64
DUTCH TOWER MILL WITH STAGE
Even the sail bars are painted in bright colours.

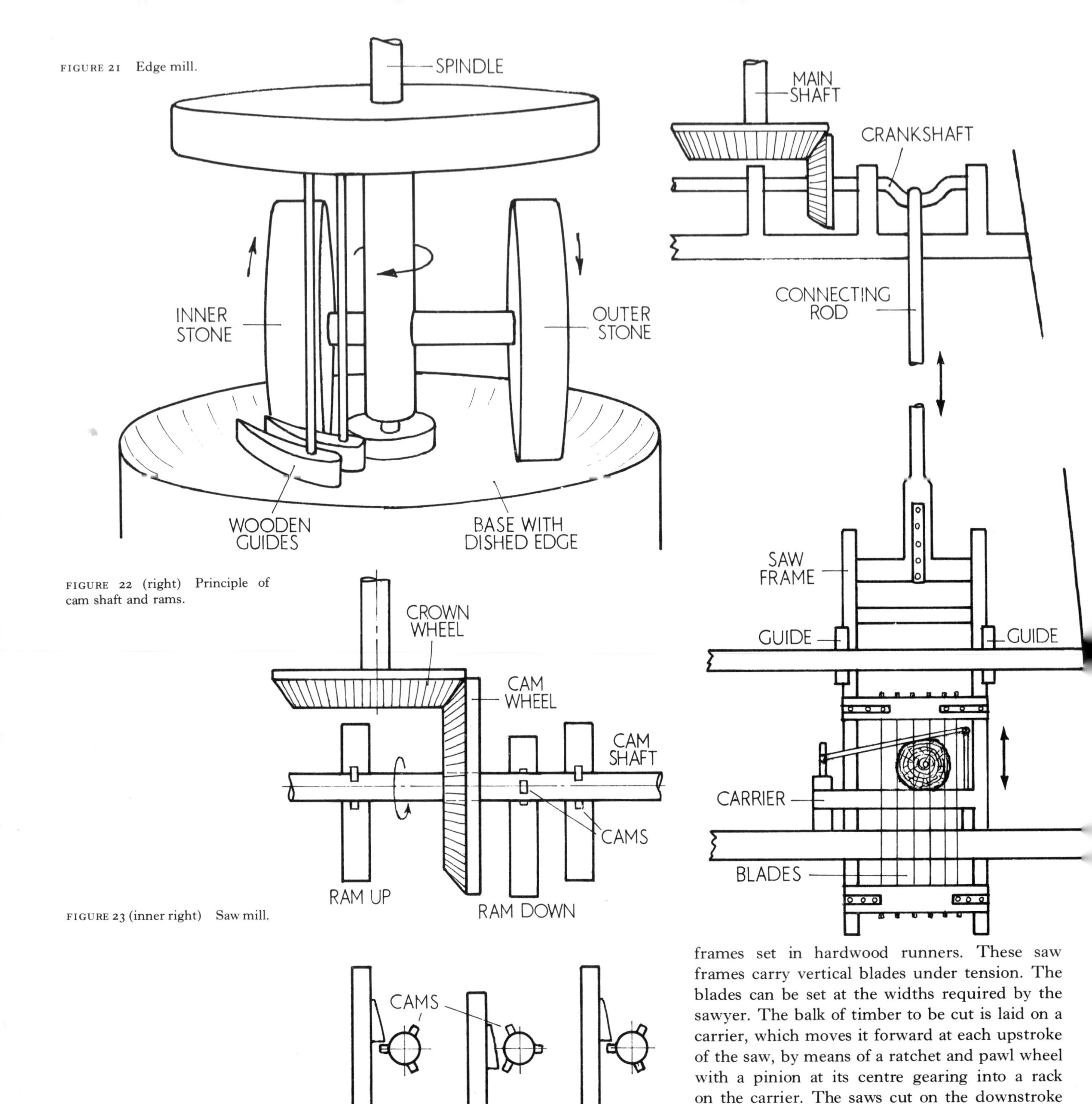

FIGURE 21 Edge mill.

FIGURE 22 (right) Principle of cam shaft and rams.

FIGURE 23 (inner right) Saw mill.

frames set in hardwood runners. These saw frames carry vertical blades under tension. The blades can be set at the widths required by the sawyer. The balk of timber to be cut is laid on a carrier, which moves it forward at each upstroke of the saw, by means of a ratchet and pawl wheel with a pinion at its centre gearing into a rack on the carrier. The saws cut on the downstroke only.

PLATE 65

DUTCH SMOCK MILL ON A BRICK BASE

A beautiful thatched mill, at Paesens in Friesland, used for grinding corn. Like most Dutch mills it has a name, 'The Dog'. The date of its building, 1861, is written in thatch behind the tailpole.

PLATE 66

DUTCH SMOCK SAWMILL, LEIDEN

This type of sawmill is capable of handling the biggest balks of timber. It was built in 1804.

PLATE 67
DUTCH PALTROK SAWMILL
'THE UNICORN', HAARLEM
The whole contraption is winded by the capstan wheel on the left.

Big smock mills are used to cut big trees. The light *paltrok* mills, which have wooden wings at their sides to accommodate long balks of timber, cut small trees or previously split timber. The paltrok mill is built of wood round a king post, with the whole body supported on a curb on the top of a circular brick wall. In principle, this is not unlike the British turret mill, except that the supporting wall is low and the body of the mill is correspondingly tall. It is winded by means of a capstan attached to a low stage at the back of the mill.

PLATE 68
DUTCH WIP MILL
A hollow post mill, or wip mill, turning under full sail in a force 4–5 breeze. It was lifting water about 4 ft from a ditch into a main dyke. The tips of the sails must have been moving at about 30 mph, and the photograph was exposed at 1/30 second in an attempt to record movement. It is noticeable on the highest sail that the tips of the sail bars are progressively more blurred towards the top where the sail was moving fastest.

Preservation Work in Holland

The Dutch have made strenuous efforts to prevent any more of their mills becoming derelict, and to restore and repair mills wherever possible. Millers who keep their mills working are subsidised by help with upkeep. Fortunately Dutch mills are so big that they can be modernised and turned into dwellings without the exterior being altered or spoiled. Other mills have become local museums, or reception centres for visitors; in Holland, some useful purpose is always found which will allow the mill to be saved. In a country where great areas would soon be flooded if pumping ceased, a failure of electric supply for any reason over any length of time could be utterly disastrous, if sufficient windmills were not available to take over. Some are therefore kept in working order at strategic points, under the provisions of the Act for the Protection of Waterways in Wartime. During World War II the windmills took over a lot of pumping work, and a great deal of grinding also went on where electric power was not available.

The Dutch Windmill Society (see Appendix 2) will send on request their information booklet, with twelve touring routes covering the best of their mills, which is invaluable if you intend to visit mills in Holland.

8

Smock and tower mills

Smock Mills

The name 'smock mill' is supposed to have originated because the mill looks like a countryman in a smock, and everyone accepts this. But to me smock mills do not look much like people in smocks. Perhaps one day the philologists will come up with another explanation of the term.

Common throughout the south and east of Britain, less common in the north-east, the Midlands and north-west, smock mills, tarred black or painted white, with white fantail and sails, look very beautiful, and, in spite of the very high costs involved nowadays, enough have been preserved to give a very good picture of the various types.

A smock mill is a many-sided wooden mill, usually on a brick base, topped by a cap which rotates on a curb and carries the sails, fantail and brake wheel. An outside stage is added to give the miller access to the sails and striking gear. One of the earliest smock mills in England, which actually bears a date, is at West Wratting on the Cambridgeshire/Suffolk border, although there were almost certainly earlier examples. Lacey Green mill dates from 1650 and was moved to its present site from Chesham in 1821. Because only the cap rotates – and not the whole body as in a post mill – a smock mill can be much bigger, taller and stronger. It has several floors, and can be built on a high brick base to lift it up to catch the wind. Union mill at Cranbrook is the best example of this being done to clear the rooftops of the town.

The brick base not only heightens the mill, but keeps the woodwork clear of water – the bottom of the *cant posts* being very vulnerable to rot. A series of these corner or cant posts, held together by *tie beams* and ledges mortised into them, make up the (usually) eight-sided frame on which the horizontal weatherboarding, or vertical tongue and groove boarding, is mounted. The tie beams support pairs of floor beams on which are laid the floor joists. The floor beams for each floor are set at right angles to those of the floor below. The stone floor usually has two extra big binder beams to take the weight of the stones and to tie the mill firmly together at that point. Between the cant posts and the tie beams are vertical studs and diagonal braces, all mortised together to make an immensely strong interlocking structure, which will withstand all the stresses of weather and moving parts upon it over decades, even centuries, of work. A windmill must remain absolutely balanced and level, and there should be no twisting or warping of the timbers beyond very small limits. The smock mill at Terling was so badly out of true that it shows in a photograph. When this happens, floors cease to be level, drive shafts go out of true, and all kinds of adjustments and makeshifts must take place before the mill will run properly.

Each cant post leans inwards from the base to the curb and each is set at an angle to its neighbour. Every mortise has to be individually cut at exactly the right angle to take the tenons on the tie beams, braces and studs. This having been done, and each 'quarter' or section between cant posts fitted together and tested for fit with each other, the eight sides would have to be brought together – the tenons pushed into the mortises, everything hammered home and dowel-pinned. One badly fitting joint could put the whole thing out, and there is a limit to how much timbers can be bent and pushed and hammered to fit without creating stresses which will eventually twist the structure or cause breakages. These mills were built long before huge cranes or tubular scaffolding was available. Sheerlegs, a tripod made of pine tree poles, with block and tackle at the top, could be used in some cases, and presumably also wooden scaffolding, but basically the mill had to be accurately made. So the millwright was a very skilful craftsman and the smock mill the supreme example of his carpentry.

The biggest disadvantage of the smock mill, as opposed to the brick tower mill, is that wood joints and boarding are not entirely waterproof. Wind and rain eventually force their way into cracks and crevices and between boards and on

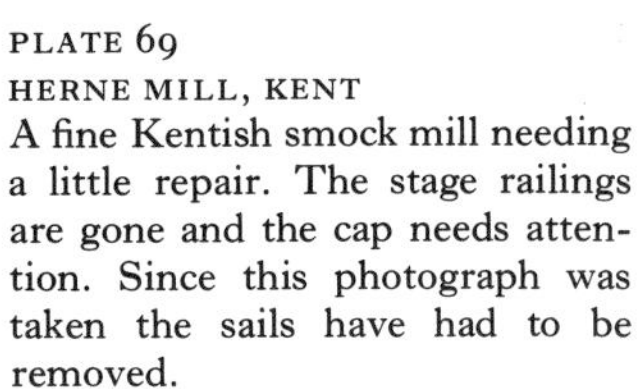

PLATE 69
HERNE MILL, KENT
A fine Kentish smock mill needing a little repair. The stage railings are gone and the cap needs attention. Since this photograph was taken the sails have had to be removed.

corners, and cause rot, and sooner or later the damage is serious. When mills are constantly at work, the miller soon spots and attends to any leaks or weak places; but the empty, still, preserved windmill can quietly turn to powder under its paint and sag without anybody noticing, until one day a big gale comes and something vital gives way. For this reason all repaired mills should be capable of turning, if not of working, and should be allowed to run as often as possible. Inspection should be much more frequent than it usually is and carried out in very bad weather when faults will show up. Ideally, where a mill is constantly under the eye of someone who knows what he is looking at, it should be possible to keep it in good condition. Never was the saying 'a stitch in time saves nine' more apt than in a smock mill.

Sometimes the inside walls of a smock mill are plastered between the studs, with rounded joins between wood and plaster, so that dust can be swept out and the mill kept clean. Lime-washed white, this plaster would also help to lighten the interior. Windows are set into each stage, but

PLATE 70 (inner left)
UNION MILL, CRANBROOK, KENT
This mill is 75 ft high, which means that the tip of the top sail is about 100 ft up. It was built in 1814, by James Humphrey, for Henry Dobell, who drank too much and went bankrupt after five years or so. Dobell's creditors took over the mill and ran it co-operatively for a while, hence its name 'Union Mill'. In 1832 it was bought by John Russell and stayed in the Russell family until taken over by Kent County Council in 1960. The fantail was added in 1840. The milling business still goes on around the mill, but except for storage uses, it works no longer.

PLATE 71
UNION MILL, CRANBROOK
Much work was done to restore the mill to its present perfect structural state. This photograph, taken while the work was in progress, shows very clearly the layout of smock mill timbering, cant posts, ledges, beams, braces, studs, etc.

PLATE 72
TERLING MILL, ESSEX
The cant posts have bent and the body of the mill is quite clearly leaning forward. Note the eight-bladed fantail stayed to the top of the ogee cap. In 1948 the miller was killed by becoming trapped in the machinery; it is believed he was leaning over the great spur wheel, oiling a bearing.

PLATE 73 (opposite page)
BARHAM MILL, KENT
This fine Kentish smock mill, which was a great landmark beside the Canterbury-Dover road, was burned down a few years ago while being repaired.

FIGURE 24 Shipley mill.

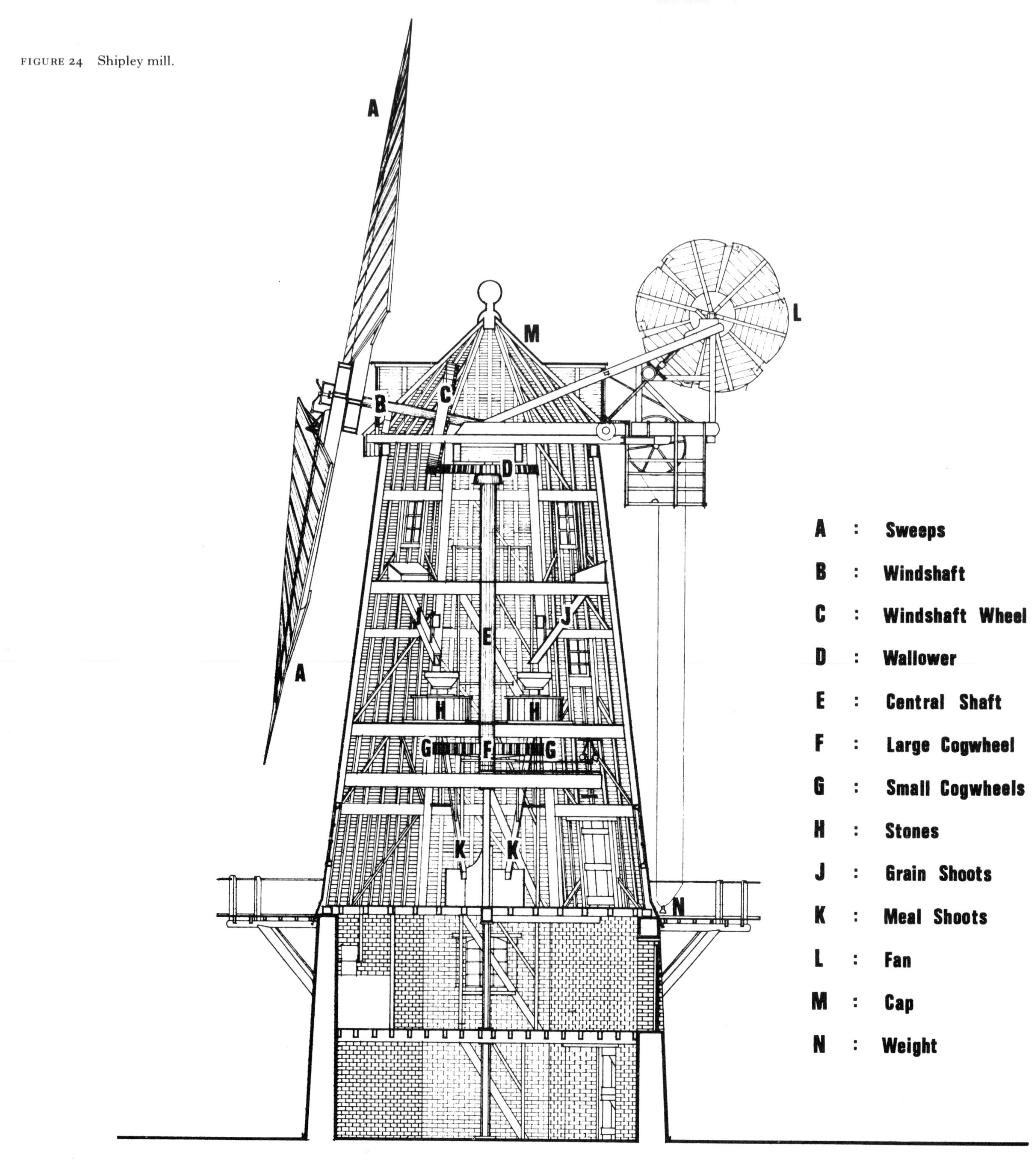

these make weak spots where the weather can force a way in, and are never very large, so the inside of the mill may be a bit gloomy.

The outside of the smock mill at Barham was covered with sheet metal on the weather side to protect it. This factor did not save it when a spark from a bonfire set it alight during repairs a few years ago. It burned like a torch and a fine mill which had withstood so many years was a heap of ashes in only a few minutes.

Dutch smock mills (plates 6, 65, 66) are thatched and, provided the thatch is looked after, are far more weatherproof than the English weatherboarded mills.

A typical smock mill has a two-storey brick base, in which grain and flour can be stored, and containing the flour dresser and bins. Near or at the top of this brick base is the outside staging or gallery, giving access to the sails (plates 69, 70). The third floor – the ground floor of the wooden part of the mill – contains horizontal shafts and pulley wheels providing the drive to subsidiary machinery (plate 61). This takes its power from a vertical counter shaft running right down from the great spur wheel. The fourth floor carries the several sets of stones, overdriven by quants from the stone nuts and great spur wheel above on the fifth floor (fig 24). The vertical drive to the great spur wheel itself comes down from the wallower and brake wheel in the cap. The fifth floor also carries the big grain bins from which the grain feeds down into the hoppers mounted on the vats (plate 26). A sack hoist at the top of the mill hauls the sacks of grain right up through a series of upward-opening trapdoors in each floor to where it can conveniently be tipped directly into the big bin. This layout varies from mill to mill. In an underdriven mill (plate 29), the stone floor would be above the great spur wheel and stone nuts, not below it, and in other mills the hoppers on the vats were filled directly from sacks, and not from a bigger bin on the floor above.

The big difference between post and smock mills is that there is much more room in a smock mill, where the various operations can be carried out and machinery housed more conveniently on separate floors. The main impression you get inside even a big post mill – such as the French mill (plate 89) – is how terribly cramped it is. There is little space in which to move around, the machinery is working dangerously close to you, and storage space is extremely limited. However, the advantage of this is that the miller can keep an eye on everything from the same place without having to climb up and down ladders from floor to floor.

Tower Mills

The internal layout of tower mills is much the same as that of smock mills, with the various machines set out on successive floors. Wide based and solid, tower mills could be built up to 80 ft high, and are therefore extremely roomy. In Britain the lower stages of such mills were occa-

PLATE 74
SUTTON PLACE, NORFOLK
The remains of one of the highest tower mills in Britain. Norfolk boat-shaped cap, and a very high stage on a grinding mill.

PLATE 75

SIX MILE HOUSE MILL, NORFOLK

Black tarred brick tower mill, on the marshes above Yarmouth. One of many which once drained this fertile grazing ground. A fine Norfolk cap with a deep petticoat.

PLATE 76

HALNAKER MILL, SUSSEX

The mill, an empty shell, has been restored by Sussex County Council more as a landmark than anything else. It has no fantail and its windows are bricked up, but its tiles still hang on the weather side. This is the 'Hannaker Mill' of Hilaire Belloc's poem:

> Sally is gone that was so kindly;
> Sally is gone from Hannaker Hill
> And the briar grows ever since then so blindly;
> And ever since then the clapper is still . . .
> And the sweeps have fallen from Hannaker Mill.
>
> Hannaker Hill is in desolation;
> Ruin a-top and a field unploughed.
> And Spirits that call on a falling nation,
> Spirits that loved her calling aloud,
> Spirits abroad in a windy cloud.
>
> Spirits that call and no one answers –
> Hannaker's down and England's done.
> Wind and thistle for pipe and dancers,
> And never a ploughman, under the sun;
> Never a ploughman, never a one.

PLATE 77
THORNTON CLEVELEYS
LANCASHIRE
A fine brick tower mill, rendered and white painted, with a Lancashire boat-shaped cap.

sionally used by the miller as living quarters, as was the more usual custom in Holland and other European countries.

Like all other domestic buildings, windmills were usually made of local materials – brick, where stone was not readily available, and stone where it was. Porous brick tower mills were frequently tarred against the weather, and some in Sussex were tile hung, at least on the side facing the prevailing south-westerlies. In Lancashire they were rendered and lime washed. While tower mills can be found all over Britain, they are outnumbered by wooden mills in the south-east, Sussex, Kent and Essex; as one moves north through Suffolk and Norfolk to Lincolnshire and through the Midlands to Lancashire, tower mills become much more common. Most of the East Anglian drainage mills were brick tower mills (plates 2, 3, 104, 108).

Some of the tallest brick tower mills, Sutton among them, have windows one above the other (plates 74, 102). Thus, between the vertical rows of windows, the brick wall soars unbroken from foundations to cap and is therefore pretty strong. Yet some millwrights – feeling that the lines of windows made weak points which could encourage vertical cracking from top to bottom of the tower – built their mills with windows in a spiral, looking out in a different direction from each floor.

The failure of a certain design can be due just as often to faulty craftsmanship or poor materials as to inherent design weakness. As far as a windmill is concerned, I feel that, if it works efficiently and does not fall to pieces or cause anything else to go wrong, then the design is good enough.

9
Living in windmills

I have mentioned that the enclosed trestle of the wip mill – the hollow post mill – provided a home for the miller and that the lower floors of the Dutch polder mills were also used as living quarters. Nowadays many polder mills in Holland have been converted into homes and weekend cottages, and the brightly painted doors and curtained windows look well under the deep thatch.

In England a lot of mills have been converted into dwellings, usually without much regard for the accuracy of the restoration; some are preserved as working structures, while others have been rebuilt and have become travesties of what they once were. However, there is something very

PLATE 78
ST MARGARET'S BAY MILL, KENT
Designed and built as an annexe to a dwelling-house. This photograph was taken in 1969 when the mill was repaired. It has recently been repainted and has a fine fantail which is lacking in this picture.

FIGURE 25 Back draught on sails.

PLATE 79 (above right)
WASHINGTON MILL, SUSSEX
It is difficult to tell this was ever a windmill at all!

attractive about the idea of living in a windmill, and lucky is the man who bought one when it could be had for a song, and made it his home.

I know of only one mill which was actually built to be lived in – at St Margaret's Bay, near Dover, in 1929. It was intended to provide additional guest accommodation for the old South Foreland lighthouse, a few hundred yards away, which with its house buildings was also occupied as an ordinary home. Architect-designed by Mr G. Lucas, FRIBA, the smock mill, weatherboarded and painted white, is built on an octagonal brickwork base, with wings which contain the living quarters. The sweeps were intended to work pumping machinery to provide irrigation for the garden, and possibly to generate electric light, although they have never been consistently used for these purposes. The mill is very near the edge of a 400 ft cliff and when the wind is off the sea there is great turbulence, and the sails will not turn smoothly – especially as there is a back-draught.

In Sussex, several mills have been converted. The one at Washington has been turned into a house, complete with television aerial on the balcony, and bears little relation to the mill it once was. Perhaps even more startling is Battle mill, which you can see for miles around, as it stands high on a hill and is whiter than whitest white. Those realistic-looking stocks and the cap are in fact made of white plastic, which will last for quite a while. From a distance at any rate, this conversion looks to the untutored eye like a real windmill. It is perhaps no worse than Rye mill, which was burnt down and rebuilt to house a bakery business; it was given a cap which is not a cap, but just a continuation of the front of the mill, and the fantail is dreadful!

High Salvington post mill above Worthing has been restored by the Sussex county council, but long before that the owner had turned the roundhouse into a café, and visitors to the mill could drink tea in a pleasant garden. Now, although the mill is preserved, the garden is not!

Weybourne mill, on the north Norfolk coast, was lived in – or at least the mill barn and house beside it was lived in – for as long as I can remember, but the mill had no sweeps and was not very impressive. Recently it has been restored and given a mock fantail, and sweeps consisting of

PLATE 80
BATTLE MILL, SUSSEX
The cap, fantail stage and windshaft are made of plastic.

PLATE 81

RYE MILL, SUSSEX

The top of the mill was rebuilt after being burned down some years ago. The cap is just an extension of the front quarter, the fantail is a twirling toy, while the stocks and whips, without sailbars, could not turn even if they had them.

PLATE 82

HIGH SALVINGTON MILL, SUSSEX

A typical Sussex post mill, worth a visit. The original owner claimed that the main post was a living tree, but when during repair work this was investigated the millwright found no traces of roots beneath the ground!

PLATE 83 (opposite page)
WEYBOURNE MILL, NORFOLK
For years the mill-house and tower were used as a dwelling, only recently have the sails and skeleton fantail been added to make it look more like a windmill.

PLATE 84
CLEY MILL, NORFOLK
The restored mill looks well. The cap is fixed and the banded fantail is not geared to it.

whips with bars bolted on to stocks. There are no shutters or striking gear and the gallery is missing, so the restoration is not complete, nevertheless the result looks splendid as you drive along the coast road.

Windmills by their very nature, being built on high places, do dominate the landscape, and yet they harmonise with it. The mill at Weybourne, as much as any mill I know, completes the scene. If you were composing a painting you would put a windmill exactly where Weybourne mill stands – in relation to the low hills and woods beyond it, and the distant salt marshes and sea.

A few miles further up the Norfolk coast, at Cley, is another fine windmill. Restored, it outwardly looks beautiful and is inwardly a house. The sweeps and fantail are not totally accurate, but look well. Two pairs of chimney pots once protruded through the cap; some years ago one pair was removed and the other painted white – rather spoiling the look of the thing.

These are just a few of the many converted windmills which exist. One could discuss for ever the pros and cons of conversions. Better that a windmill should be made to look like the ones at Weybourne or Cley – an asset to the landscape – than that it should disappear altogether or exist as at Rye, Washington and Battle. Best of all it should be restored exactly as it was, but the enormous cost precludes this in most cases.

10

French windmills

France has 3,000 mills, the oldest dated 1183 – more than England, as many as Portugal – but unfortunately fewer than fifty are in working order and many are in a semi-derelict condition. Nevertheless, there is an active Association of Friends of the Windmill, and really strenuous efforts are being made to carry out the work of rescue and repair when the cash can be raised to do it. The first priority is to save the wooden mills, of which only about forty remain; then to preserve one of each type of mill in each region, and at the same time to concentrate on those which will be of

PLATE 85
FRENCH TOWER MILL
Restored as part of a house, it is typical of hundreds of its type.

PLATE 86 (opposite page)
PETIT PIED TOWER MILL
BRITTANY
These little mills are very common in parts of Brittany and have a pleasing, well-balanced shape. It would make a splendid, if rather small, country cottage.

immediate interest to tourists. Quite a programme!

In a recent bulletin published by the association, the editorial – translated here – seemed to me to sum up very conclusively the answers to the question so often asked: 'Why save the windmills?'

Because men want and wish to know their origins and how their ancestors lived.

Mills had a preponderant place in simple everyday life. They were in effect the objects of many conflicts of right and gave rise to many expressions still used in our times.

Because men are curious about old techniques; think for a moment that these 'machines' have not for all practical purposes changed during 600 years – 25 generations.

Because men can once again by using mills utilise two natural sources of energy: water and wind, without using them up and without pollution.

Because men have converted with courage and love these magnificent edifices which in each region of France blend so well with the countryside.

Because men need once again the help of the windmills, to make flour and oil. To saw and turn wood, to grind and mix various products, to pump water for irrigation, to give life to fountains, and many other possible practical uses.

Because men should pass on their heritage to their children and to the men of tomorrow.

Enough said!

There are various types of windmill in France, from the post mills of Flanders – basically the same as those of Belgium and Holland – to the 'Parisien' type of tall post mill with a tiled roof, not unlike the Dutch wip mill in shape. Tower mills – perfectly cylindrical with conical caps – are to be found all over central France. The windshafts are horizontal or only very slightly angled upwards. Winding of these tower mills is almost always done by a tail beam, with perhaps one brace, coming down from the back of the cap. The caps are tiled or boarded. In Brittany one finds the very attractive tower mills known as *petits pieds* – 'little feet' – so called because they taper, being narrower at the bottom than at the top. Very beautifully built, with a niche for a statue, and decorative balconies and courses, these little mills have lasted well and many are being preserved.

The sails on the older tower mills are weathered, but consist of sail bars on long stocks, protruding an equal distance on both the leading and trailing edge, and without hemlaths. Cloth sails would be laced and tied to these bars (plate 16).

The tower mill in plate 87 has longitudinal shutters which appear to be worked by a striking mechanism. They move longitudinally like a multiple parallel ruler and, when shut and not in use, form a long narrow bar.

PLATE 87 (above)
FRENCH TOWER MILL
The longitudinal shutters close up like the bars of a parallel ruler.

PLATE 88 (above right)
'MICHEL', COQUELLES, CALAIS
The enormous ladder suspended from the buck is supported by the massive tailpole which also carries the winding gear. The brake line is hanging down below the buck. Notice how the 200-year-old tailpole has dried and split.

At Coquelles – not far from where the Channel Tunnel was to have spewed out its trains, cars and people – is that European rarity, a working windmill. This pre-Revolution wooden mill, named 'Michel', has stood up to a couple of centuries of north European weather and survived several wars. Without any back-up engine power, the mill works whenever the wind allows, to produce various types of flour for health food shops, and does a steady trade. It is a very big open post mill; to stand and look down its ladder from the doorway of the buck is quite startling.

Its unpainted timbers have weathered and shrunk, but it remains perfectly balanced, runs efficiently and quietly, and – except for a few iron clamps and some corrugated iron sheeting to protect it on the weather side – is little changed from the day it was built. It has two sets of stones, very big by English standards: one driven directly

PLATE 89
'MICHEL', COQUELLES, CALAIS
The side balcony of this open post mill houses the flour dressers.

from the brake wheel (plate 25) and another directly from a tail wheel at the other end of the windshaft. Flour-dressing machinery is housed in the gallery at one side. Winding is done by a windlass, crank and a wire hawser which hooks on to rings set in the ground (plate 33). The whole weight of the ladder, tailpole and windlass hangs from the back of the mill and is unsupported at ground level, balancing the big sails and heavy machinery in the buck (plate 88).

The mill's colourful history, in this literal translation from the French, speaks for itself:

It is one of the most ancient mills of Flanders. It dates from the beginning of the eighteenth century. In 1804 an exceptionally strong gale blew it down, as it blew down the seventeen mills of Mont Cassel. In 1805, put back in place, it continued its service, crushing the good grain from the wheat, flattening the secondary cereals for animal feed, under the surveillance of the millers of Coquelles of whom the last is Master André Darré. On 22 May 1940, Hitler's armies arrived near Calais. The mill and the bell tower of the church of Coquelles were admirable guiding marks for the Royal Navy who conscientiously threw their shells on Route Nationale 1. The Nazis saw this and blew up the old mill. On his return from the wars, Master Darré took account of the disaster. His life was changed, because a miller lives by his mill, in his mill and for his mill.

At the Liberation, ill but not discouraged, the miller and his friend, the carpenter Eugene Roos, and the town clerk, Michel Grassien, go to find a mill for sale in Flanders.

They find the mill of Arneke, the oldest in Europe, which dates from 1001 [sic]. But its owner wants too much money and does not really want to part with the mill. They discover the mill of Crochete. It belongs to Michel Deconninck, whose father, the miller, is dead. He agrees to sell the mill which grinds good wheat flour. We are in 1950. In 1952, thanks to the perseverance of the miller of Coquelles and of the two carpenters, Eugene Roos and Achille Lejeune, and with the agreement of the town clerk who backs them up with the services of the Ministry of Reconstruction, the mill of Coquelles turns again to every wind.

It is exactly the brother of the original, carrying the date 1757 and the *fleur de lys* of the king of France. The brake wheel carries the date 1776. The weather cock is on the peak of the roof. Linen cloths cover the sails and the natural silk filter in the flour dresser, separates the flour from the husk and bran.

A local poet writes the *Hymn of my Miller*. But there is a shadow on the picture. The same day that the work of reconstruction finishes, Achille Lejeune is sawing through the floor to make a hole for a trapdoor. To break the slab free, he hits it with a crowbar. But, carried by the movement, he goes with it through the gap and falls seven metres. He breaks his cervical vertebra on one of the beams. The mill is never to be inaugurated, but it is baptised as is customary, by the village priest. It is called Michel; its godfather is the Mayor M. Mobailly, its godmother is Madame Grassien, wife of the town clerk.

The mill weighs 60 tons, develops 60hp in normal weather, and the speed of the sails can be reduced by reefing the canvas.

The miller is always between sky and earth. He can, thanks to constant observation, foretell the weather. He is happy in his mill; he is at ease and he is never cold.

The mill is alive, it has a soul, and with it one dare not cheat. Its flour is pure, and is not mixed with other ingredients. Nowadays shops which specialise in health foods have contracts with Master Darré for the supply of good natural flour. . .

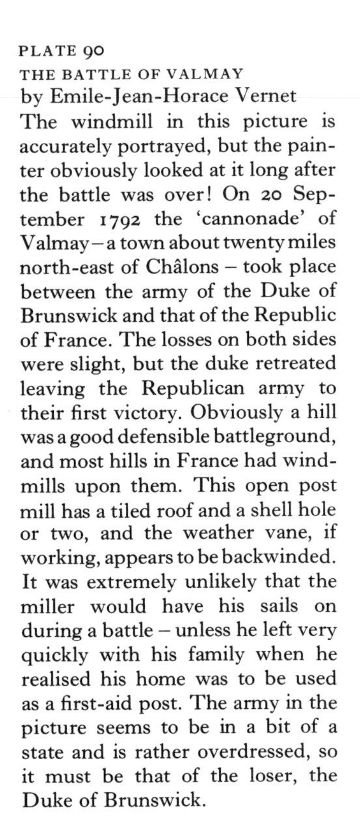

PLATE 90
THE BATTLE OF VALMAY
by Emile-Jean-Horace Vernet
The windmill in this picture is accurately portrayed, but the painter obviously looked at it long after the battle was over! On 20 September 1792 the 'cannonade' of Valmay – a town about twenty miles north-east of Châlons – took place between the army of the Duke of Brunswick and that of the Republic of France. The losses on both sides were slight, but the duke retreated leaving the Republican army to their first victory. Obviously a hill was a good defensible battleground, and most hills in France had windmills upon them. This open post mill has a tiled roof and a shell hole or two, and the weather vane, if working, appears to be backwinded. It was extremely unlikely that the miller would have his sails on during a battle – unless he left very quickly with his family when he realised his home was to be used as a first-aid post. The army in the picture seems to be in a bit of a state and is rather overdressed, so it must be that of the loser, the Duke of Brunswick.

11
Caps and curbs

The cap of a tower mill or smock mill has to be immensely strong; it carries all the weight and stresses of the sails, windshaft, brake wheel and fantail, and at the same time has to be free moving on its curb. It also keeps the top of the mill weatherproof.

The cap must be so shaped that it allows the air flowing past the sails to run cleanly away without eddying or back winding. So far as I know, no full-scale wind-tunnel tests have ever been carried out on windmill caps – in any case, such an exercise would be purely academic as windmill design is not being developed or improved. Cap shape tends to vary from one locality to another; although theoretically some shapes should be better than others, in practice – provided a mill develops plenty of power and runs smoothly over a wide variety of conditions – who is to say that its cap design is bad?

In south-east England, particularly in Kent, the smock mills have rather bluff caps with straight ridges. These are covered with weatherboarding and are really a development of the shape of the top of a post mill. This shape, also found in parts of Sussex and Essex, should, aerodynamically, be bad, but hundreds of mills in these areas worked perfectly well with it. In Sussex the ogee shape is popular; this is a round domed cap with slight upward reverse curve to a *finial* knob (plate 76). The *beehive* – the Kentish name for a domed cap of any type – has a *petticoat* which protects the curb and the junction between cap and body of the mill from the weather. Sometimes, as at Halnaker mill, the petticoat is covered with sheeting. Usually, as at Shipley or Punnetts Town, in Sussex, it shows as vertical boarding; these two are interesting variations, somewhere between dome and ogee.

PLATE 91
HERNE MILL, KENT
Typical Kentish cap. The cap frame can be seen under the weatherboarding. The cogs on the curb engage with a worm drive which is just visible at the back.

PLATE 92

SHIPLEY MILL, SUSSEX

The youngest and biggest smock mill in Sussex, beautifully restored by the County Council, is variously known as Vincent's mill, Belloc's mill, King's mill, or even as Mrs Shipley. It has an unusual pyramid ogee cap with a knob finial and a dormer fore and aft giving access to the sweeps and fantail stage. The vertically boarded petticoat is neatly scalloped. Built in 1879 by Mr Grist, millwright, of Horsham – who sounds as if he should be in 'Happy Families' – it cost £2,500. In 1958 repairs to sweeps, flooring and a few outside elements cost £4,556 – and today they would cost almost twice as much. In 1906 the mill was bought by Hilaire Belloc, and now carries this memorial plaque:

Let this be a memorial to
Hillaire Belloc
Who garnered a harvest
of wisdom and sympathy
for young and old
MDCCCLXXX–MCMLIII

PLATE 93
BLACKDOWN MILL, PUNNETTS TOWN, SUSSEX
This ogee cap with an iron gallery frame and boarded petticoat was designed by the miller. He covered the cap with metal sheets from an old bus.

The ogee cap is also found in the East Midlands and the north-east of England. The Cambridgeshire and Lincolnshire caps are much more shapely than the Sussex ogees, and have much finer and taller finials with acorn-shaped knobs. In the same area is found the *pepper pot* cap, a dome without reverse curve and finial, but with a hatch in its flat top.

Dome-shaped caps (plates 18, 114), can be seen almost anywhere; they are usually covered with sheet metal – nowadays aluminium. While the ogee has a concave curve upwards towards the finial, the dome is convex all over, but either shape can be a compromise according to the millwright who built it. Some fine ogee caps actually have a reverse curve at the bottom, making them almost onion shaped, and curve right round, dispensing with a vertically-boarded petticoat.

The boat-shaped cap of the Norfolk Broads is, to my eye, the most attractive of all. The millwrights, Englands of Ludham, claimed to have introduced the shape, which is really very like a clinker-built dinghy, with the same curving lines. Anyone used to boats would feel that a shape which goes through water will also allow air to flow evenly along it; and, by turning upside down a boat built to be waterproof, and setting the boards the other way round, he could make an equally waterproof cap (plates 2, 3, 31, 104, 108). The curves of the ridge and sides were harder to build than those of the Kentish cap. A Norfolk cap usually has a gallery round it to give access to the sails, fantail and the roof of the cap itself for repair and painting. Most of these caps have vertical board petticoats. It would not be necessary to climb out on the ridge itself to tar the cap of a Norfolk mill.

Another distinctive but rare type is the *pent roof cap*, in which the eaved roof is built sloping directly up from the outer edge of the basic frame to a straight or sloping ridge, looking rather like the roof of a house, with the windshaft coming out through the gable end.

The Lancashire boat-shaped cap is bigger than the Norfolk version and, seen from above, widens out very noticeably to take the top of the tower. It is gabled at both ends, but narrower at the tail than the head, almost coffin-shaped (plates 36, 77).

Within these six main British types of cap –

PLATE 94 (opposite page)
CATTELLS MILL, WILLINGHAM
CAMBRIDGESHIRE
Typical Cambridgeshire shape with a high brick base and a weatherboarded smock top. The cap has a fine pointed finial and knob. This mill, with double sided sails, must have been turning when the photograph was taken, as its shutters are half closed.

PLATE 95
HERNE MILL, KENT
Tarring the cap of Herne mill. The millwright is holding on with his knees and must have the bucket of tar in front of him. Presumably he worked backwards and eventually climbed back on to the fan stage. The photographer had climbed high on the fantail beams to get his picture, which also shows very clearly the fitting of sail backs and bars to whips and stocks.

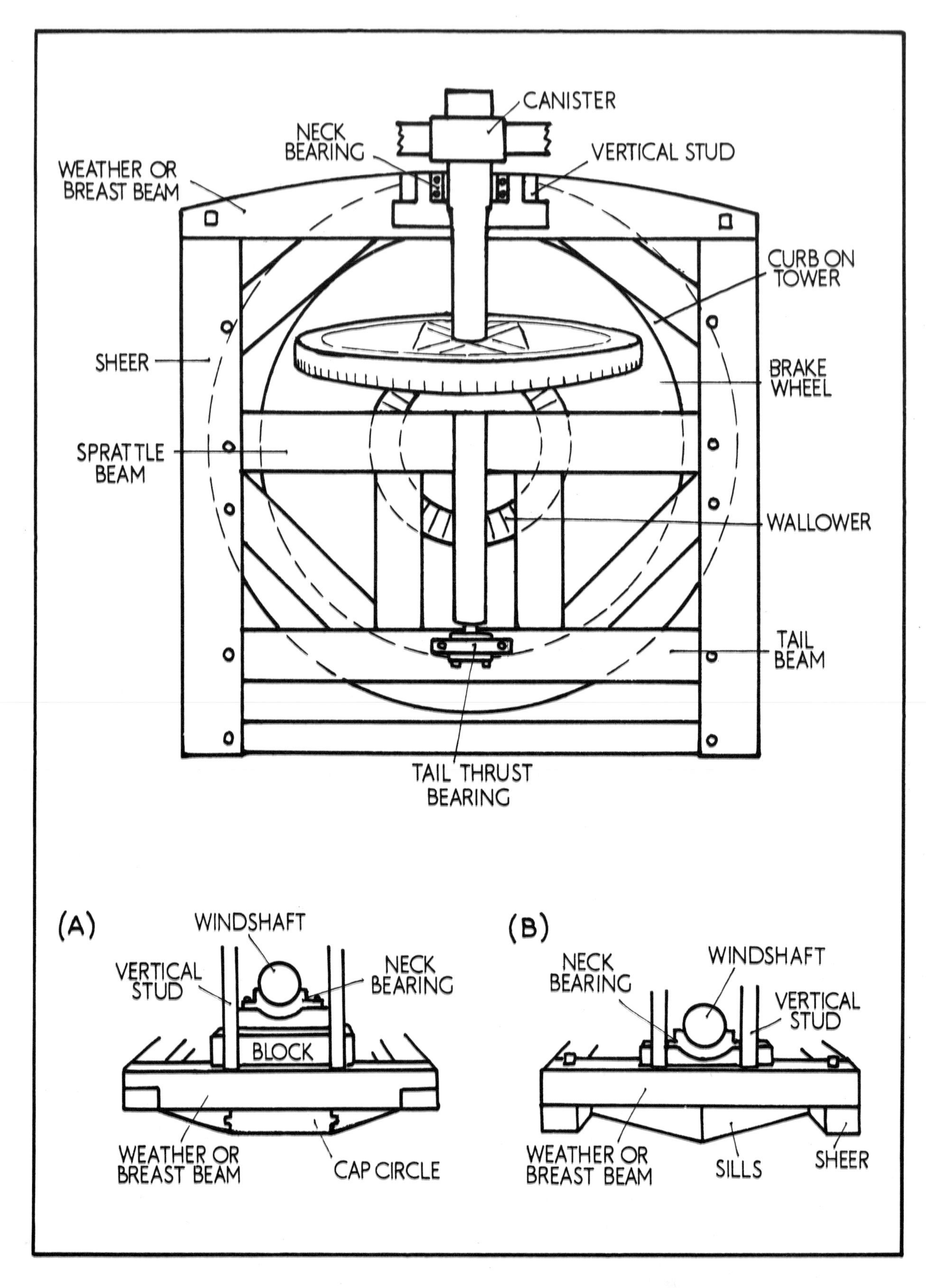

FIGURE 26 Cap frame designs: (A) cap frame with block above and cap circle below (B) cap frame with sills.

Norfolk boat shaped, Kentish bluff, Sussex and East Midland ogee, dome shaped, pent roofed, and Lancashire boat shaped – there are many variations and compromises. The basic framework normally consists of two main beams or *sheers* with tie beams between them. The tie beam at the front, the *breast beam*, carries the *neck bearing* and its retaining *vertical neck studs* through which the windshaft passes into the cap. The centre or *sprattle beam* carries the top bearing of the upright shaft. The tail beam across the back end carries the *thrust bearing* to take the end of the windshaft. In some types the sheers project beyond the end of the cap to take the *fan staging*, which is otherwise made up of smaller beams tying into the inside of the cap. This basic oblong framework must turn on the curb, so it either has a *cap circle* of up to eight pieces fixed below it, or it may itself incorporate diagonal pieces or *cap sills*, which not only brace the frame but complete a circle with the sheers and tie beams. In some cases there is no complete circle at all. *Rollers*, or blocks, or a track are fitted on the lower surface of this cap frame. These turn on the curb on the top of the tower. There are different types of curb, but whatever the type it must allow the cap to turn completely freely. Subsidence of a tower causes breakage of the curb and consequent jamming or stiff movement, which will soon lead to serious damage or even to the destruction of the mill. Some mills have wedges fitted with which to adjust the curb level.

According to the type of winding gear, the curb must have cogs on it which engage with a screw or worm or pinion wheel, actuated by a hand winding wheel or by the fantail mechanism. As the worm or pinion moves one way or the other, it passes the cogs along itself and turns the cap. This cog and worm arrangement can be inside or outside the mill. A mill which is winded solely by means of a tailpole does not need to have this bearing as the cap only has to rotate on its curb when it is pushed or pulled around (plates 31, 103).

The cap must be carefully centred on the tower; this is done by mounting four *centering wheels* on the cap frame which run against the inside of the curb. Sometimes there is a flange above these wheels which helps also to keep the cap down on the tower should it be backwinded or otherwise unbalanced. In other cases iron hooks, known as *keep irons*, come down over the flange to do this job.

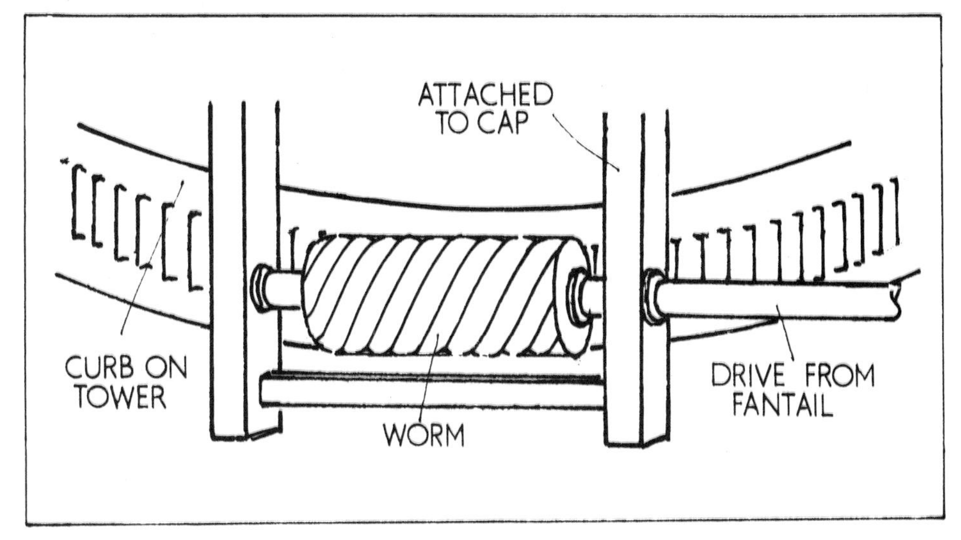

FIGURE 27 Worm and cap cogs.

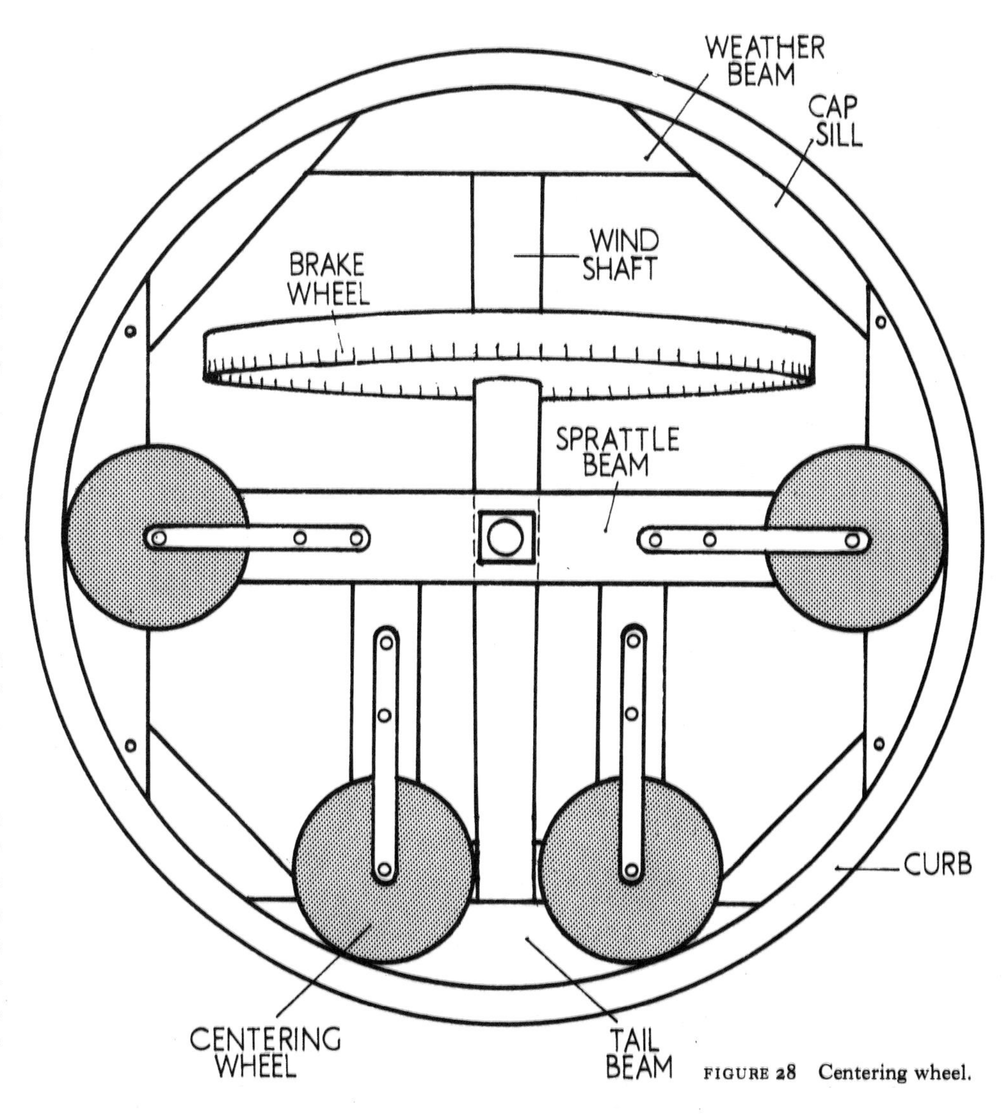

FIGURE 28 Centering wheel.

PLATE 96
EASTRY MILL, KENT
Looking up from below at the sprattle beam, brake wheel (without cogs) and the top bearing of the upright shaft – the shaft, with the wallower, has been removed. One of the centering wheels and the straps which hold the other three can be seen clearly, as can the iron brake going right round the brake wheel.

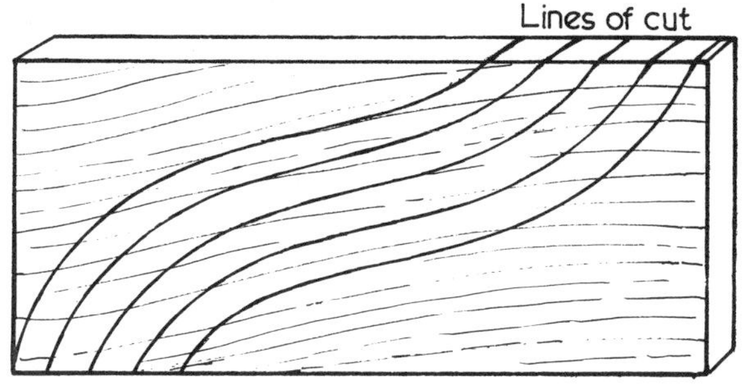

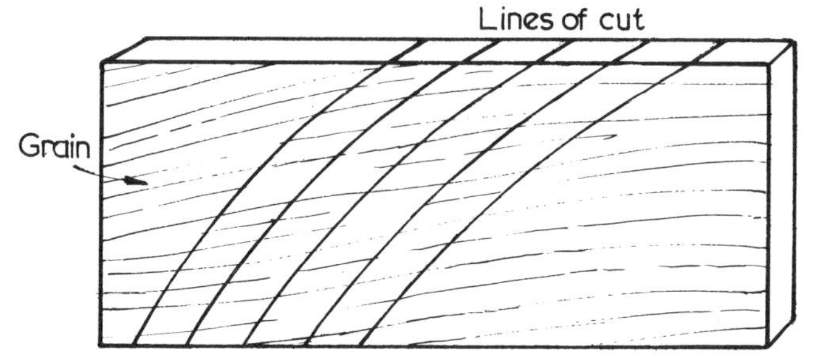

FIGURE 29 Cutting cap ribs.

The roof of the cap is built up on the basic framework, described above, on ribs or rafters of oak, elm or poplar. Cap ribs are cut on the curve from flat planks, following the curved grain of the wood wherever possible; therefore the width of the planks available determines the fullness of the curves. Ribs for extreme curves can be cut in sections and scarfed together, but this makes a slightly weaker job. Sometimes, in the old days, poplar ribs were steamed to shape, or timber grown with a curve cut so that the grain ran true along the rib.

The ribs must have the shape of the cap. In the Norfolk caps, they had to be made rather like the ribs of a boat, each one individually tailored (plate 31). Wicken Fen mill has most pronounced curving – although the board covering is vertical and not the usual horizontal (plate 103). The bluff

PLATE 97
HERNE MILL, KENT
The cap of the mill being repaired. This shows very clearly how the cap ribs are set up.

Kentish caps, ogees and dome-shaped caps required identically curved spars. Overlapping weatherboarding, or butted boarding covered with sheet metal or painted canvas, complete the job.

The cap of Blackdown mill at Punnetts Town, in Sussex, was designed and made by the miller and sheeted with metal from an old bus! (plate 93).

Thatched caps were once common in Somerset, but only one now remains – on Stembridge mill at High Ham.

The caps of Dutch tower and smock mills are often thatched; they are to some extent boat shaped, but taper off even more sharply towards the back than do the English ones. This is to make a shape on which thatch can conveniently be laid to carry away water (plate 20). The beams which carry the tailpole and braces are an integral part of the cap framework.

The conical cap of the Spanish pumping mill (plate 110) is beautifully simple and in a dry climate was practical enough.

12

Protection and preservation

PLATE 98
OUTWOOD MILL, SURREY
Dying – the last stage of decay. This mill has a neighbour, a post mill, which has been repaired.

Time, weather and neglect are the biggest enemies of windmills, and it is surprising how quickly, once the rot sets in, damage can be done which will cost thousands of pounds to repair. Mills need protection from the depredations of man himself – not always vandalism, often sheer ignorance or lack of interest. A mill in private ownership may come up for sale, and be bought and demolished before anyone realises what is going on. It needs continuous vigilance and inspection to counteract these destructive processes.

A mill which may have been perfectly sound when you last visited it a year ago may have lost half a sweep in the last gale, or be showing signs of rot where rain is forcing its way through joins in weatherboarding. Mills which are not in use sag and open up because, instead of continual variations of the type and direction of stresses which cancel each other out in a working mill, the stresses remain constant in the same places. There are just not enough experts to go round to enable a constant regular check to be kept on all mills, so it is up to you, even if you do not have expert knowledge, to do what you can to record and report any changes in the mills that you visit. These remarks do not, of course, apply to those repaired and preserved mills in public or private ownership which have a resident miller or custodian.

Mills in private ownership present a great problem because there is little anyone can do to stop the rot if the owner is not co-operative. Additionally there are still large areas of the country where windmills and windmill remains have not been fully recorded, and once you have a little experience and knowledge there is fascinating work to be done in this respect.

Windmill Societies

The Society for the Protection of Ancient Buildings is exactly what its title says it is, and that includes mills. Anyone seriously interested in windmills should join its thriving Wind and Watermill Section, which co-ordinates records and supervises work generally. It organises lectures and windmill tours, and works with and through the handful of millwrights and millwrighting firms still active in Britain to give free advice to members and others about their mills. It prepares reports and raises very substantial sums of money to be given as grants towards mill repair, and advises county councils, the Ministry of Works and other public bodies on windmill repair and upkeep. It also provides record cards for members wishing to measure and record windmills. The secretary of the Wind and Watermill Section of SPAB, Mrs Monica Dance, together with a voluntary committee which includes many of the top professional and amateur windmill experts in the country, does an enormous amount of work, but funds are always needed and new members always welcome (see Appendix 2).

Apart from this national organisation, there are local windmill societies. Quite a few mills have been saved because local people got together to raise money and find expert help (usually through SPAB) to do something about a windmill which was falling down. If there is such a mill in your neighbourhood and nothing seems to be being done about it, you might, if you are energetic enough, be able to organise a society to save it.

The first step is to find out who owns the mill and to work from there; here again SPAB will help you. Do not be discouraged if at first the owners of the mill send you packing. It is usually because they know that repairs will cost a great deal of money and are frightened of getting involved. Nevertheless expert advice and the promise of funds can do a lot to change people's minds. In many cases – while it may not be possible to do all the repairs necessary to restore the mill to what it was – enough may be done to prevent it from falling down, to hold it for a year or two while more funds are raised and more interest engendered.

A case in point is Sandwich mill. In 1960 this mill was in a bad condition. It had not worked by

NAME AND ADDRESS OF MILL			COUNTY	REF.No.
Nat.Grid Reference or Location	Industry	Dating	Parish or Township	Date of Report

DESCRIPTION	WHEEL OR SAILS
CONDITION	

Danger of Demolition or Damage

Printed, Manuscript or Photographic Records

Reported by	Owner of Mill
Return Card to	Tenant of Mill

S.P.A.B. Wind & Watermill Section Watermill Record Card

MACHINERY DETAILS	
DRIVE SYSTEM	MILL MACHINERY
AUXILIARY MACHINERY	REMARKS etc.

FIGURE 30
SPAB record cards.

PLATE 99
Dead – the bones of Outwood mill, Surrey.

wind since 1926 although, linked to a diesel engine, it had gone on grinding up to 1954. Vincent Pargeter – then a bank clerk, now a professional millwright, working in his spare time and initially with his own money – renewed some of the cant posts and braces and studding, put on weatherboarding where necessary, and made and fitted a pair of sails. Now, twelve years later, he is again working on this mill. The white paint he put on before needs renewing and she looks a little shabby, but the second pair of sails will be in position before long, and the mill restored both internally and externally.

The money for this work is being provided by Sandwich Borough Council, who have realised that the mill is an asset to the beautiful little town. Had Mr Pargeter not saved the mill in the first place it might by now have become totally beyond repair.

A new cap has recently been put on Eastry mill,

PLATE 100
SANDWICH MILL, KENT
Putting on new sails.

in Kent (plate 96). This mill is privately owned and, much as the owner would like to carry out more work on it, it is unlikely that this can be done, for the time being at any rate. But the mill body is in reasonable condition and, although repairs are needed, it should not now deteriorate.

Many county and district councils give financial help with windmill repairs. In Norfolk, where there are numerous windmills, the county council has provided funds and the impetus to restore several of these, and will do more when the money is available. In Essex the council has engaged a full-time millwright to work on the restoration of mills in the county.

PLATE 101
BARHAM MILL, KENT
Although it was burned down, this mill may yet be rebuilt. To recreate it from scratch would be a costly job.

13

Drainage mills

In East Anglia, drainage mills were used, as in Holland, to reclaim marshland and to keep it free of water. A Dutchman, Vermuyden, came to England in the reign of Charles I to drain the Fens and teach the English the art which had been perfected in Holland while the Fens were still malarial swamps. Windmills, the only source of power for pumping water, were used until 1820 when the first steam pump was installed on Ten Mile Bank. J. A. Clark, in *Fen Sketches* (1852), estimated that there must have been more than 700 windmills 'between Lincoln and Cambridge'; when he wrote these had been reduced to about fifty mills in the Bedford Level, or a total of 220. In 1838 another writer, J. Glynn, stated that 'in 1830 Littleport Fen, about 28,000 acres, were drained by two steam engines of 30 and 80 horse power with a few old wind engines still retained. Before steam power was used there were over seventy-five windmills in this district'.

The Fenland mills have disappeared completely. I lived in Spalding as a child and I cannot remember a single drainage mill. If I think of mills in Lincolnshire, they are the great tower mills at Boston, Alford and Long Sutton, which were used for grinding not pumping. The Fens in clear light are magnificent. The sky is all the picture – the flat land, gleaming black plough or green grass, only the lower fifth of it. The eye finds the scale from distant church spires or a line of lonely poplars. A great windy overwhelming landscape it is. But how wonderful it must have been when everywhere you looked there were windmills, rows of turning windmills, bringing the whole landscape to life and giving man and his works a place in the scheme of things. No wonder that part of Lincolnshire is called 'Holland' for when the windmills were there it must have looked exactly like it.

Norfolk, on the other hand, waited another hundred years before it turned over to steam pumps. Perhaps the drainage authorities had less money, but in any case the area of the Broads drained by windmills into the river system is much smaller in total than that of the Fens. In fact, little mill draining was done until the end of the eighteenth and beginning of the nineteenth century. The area in prehistoric times was a complex of rivers and estuaries. The land rose slowly, and the rivers flowed out through peat marshes. The vegetation here eventually formed peat beds, and the whole area rose clear of the sea level, so that up to Romano-British times it consisted of rivers running through low-lying peat bogs to the sea, subject to occasional incursions of the sea. Until the end of the thirteenth century, the land was 10 ft higher than it is now, and it has slowly subsided again ever since.

The Broads themselves were formed by peat-cutting beside the rivers in the fourteenth and fifteenth centuries. Since then the land has been used largely for grazing, but by the late 1700s the general subsidence made it necessary for a lot of drainage work to be carried out. Ditches and dykes were dug by hand, and windmills installed at the junctions of dykes and rivers. One windmill might cope with a 'level' of 1,000 acres, but it was not necessary to have as many windmills together as in the Fens or in Holland. The particular level and all the mills draining it were the responsibility of the marshmen, as was the stock grazing there. Marshmen lived in or beside the mills and sometimes had more than one to look after.

The Fenland drainage mills were mostly smock mills, some of them quite small. The little mill at Wicken Fen, with its common sails and tailpole winding gear, is a typical example. It is situated in a nature reserve, and was moved there from nearby Adventurers Fen. The Broadland drainage mills were mostly brick tower mills (plate 3) with wooden caps. Some had straight-sided solid looking towers (plates 2, 3, 75), while others were slightly waisted (plate 2) and very elegant. The mill at Thurne Dyke, now restored and used as a windmill museum, has this waisted shape. I have a pastel drawing of her done by my grandfather; we always knew her as the 'Lady Mill' because she looked like a lady in a high-waisted, long-skirted dress. Perhaps the most typical part of

PLATE 102

ALFORD MILL, LINCOLNSHIRE

A fine five-sailer corn mill. The ogee cap has a slight reverse curve at the bottom, making it almost onion shaped and fitting it very snugly to the top of the tower. The sails have no boards or shutters on the leading edges, a common feature of multi-sailed mills.

PLATE 103

WICKEN FEN, CAMBRIDGESHIRE

Small preserved drainage mill with an inside scoop wheel. Common sails, two of which have been set for the benefit of the group of windmill enthusiasts visiting the mill. Note the tailpole and braces for winding the mill.

PLATE 104

THURNE DYKE MILL, NORFOLK

Carefully restored, the mill had just lost its fantail in a storm; this has now been replaced. It is well known to all who sail the river Bure, but it is a pity it has no shutters or striking gear so can never turn. It was backwinded in 1919 and the whole of the cap, sails and a lot of machinery ended up in the marsh. Luckily for us the mill was repaired.

PLATE 105

HOW HILL MILL, RIVER ANT NORFOLK

Only the trestle now remains. In this old photograph the mill is rather out of focus, as the photographer was concentrating on the wherry. These Broadland barges carried grain and other goods to and from the mills and Broadland villages right up to the 1930s. One or two survive as pleasure boats.

the Norfolk drainage mill is the boat-shaped cap, which can be seen in all these illustrations.

In addition to the tower mills, there were small trestle drainage mills. The trestle of the one at How Hill on the river Ant remains, as does another at Horning. These small trestle mills carried a windshaft, sails and fantail on an open work trestle, with an iron crown wheel and pit wheel turning a small iron scoop wheel fitted with wooden scoops, set in a dyke between the feet of the trestle.

The Norfolk draining mill contains little machinery. A wallower transmits power down a vertical shaft to a bevel crown wheel, which engages a big pit wheel on a horizontal axle. This wheel runs in a pit or trench dug in the floor of the mill. The other end of the axle carries the scoop wheel which could be located inside the mill (plate 103), just outside it, or detached from it – as at Berney Arms, where the scoop in its casing stood quite separate from the mill, driven by a long axle (plate 7). The scoop wheel, turning in a trough with bricked-in sides and bottom just slightly bigger than the wheel, pushes the water up about 5ft and through an iron sluice gate into the tail race, which discharges into the river. The sluice gate is opened automatically by the pressure of the water being thrown against it, but shuts tight if the river water rises to push against it at high tide or in flood conditions.

At normal working speed a scoop wheel could lift something like seven or eight tons of water a minute; this may seem quite a lot, but is only a fraction of the output of modern electric pumps. Nevertheless, the output was sufficient to cope with almost all circumstances, except really bad floods. Windmills were abandoned not because they did not do the job, but because they became expensive to keep in repair, because the wages of the marshmen had to go up, because the mills could not be switched on and off exactly when needed, and because one pump could do the work of several windmills.

The steam pump ousted the windmill, and the electric pump has ousted the steam pump. The Dutch are farsighted enough to have kept some windmills for use in emergency, but the drainage of the Broads could get into a pretty good mess if electricity supplies failed for any length of time. There could come a day when all those windmills, which twirled so merrily when I was a child in the 1920s and early 1930s, will be sorely missed.

PLATE 106
HORNING TRESTLE MILL
NORFOLK
The crown wheel, pit wheel and scoop wheel between the legs of Horning mill, which is of the same type as the one at How Hill.

PLATE 107
NORFOLK MILL
Main upright shaft, crown wheel and pit wheel of a typical drainage mill.

PLATE 109
STARSTON MILL, HARLESTON
NORFOLK
An oddity, this is a little iron post mill with wooden sails. It winded itself like a weather vane and was used solely to pump water for a garden.

PLATE 108
LOCK GATE MILL, BREYDON
NORFOLK
Wooden scoop wheel on a metal frame. Note the Norfolk boat-shaped cap and decorative window frame and fanlight.

14

Windmills in America and elsewhere

Many mills were built in America by the colonists and persisted until the advent of the steam engine. It is estimated that at one time there were 7,500 mills in New York State, a high proportion of which were watermills. In relation to the size of the country, a large number of windmills must have been concentrated on its eastern seaboard, but few remain. Most of these have been turned into small museums or set up as part of rural life parks, and some have been converted into residences.

In any area in the world where running water is plentiful, watermills, which can operate continuously as long as there is a water supply, were always built in preference to windmills. The problems of shape and size which arise with windmills do not affect the construction of watermills, which are really wooden or brick factories or workshops located beside or over streams, taking their power by gearing from the axle of the water-driven paddle wheel. Windmills were built only where the water supply was not sufficient to turn a mill.

Windmills were being built in America at a time when the European post mill was already fully developed. The construction of this extremely efficient type of mill must have been familiar to the immigrants from Europe in the eighteenth century. No original and complete post mill remains in America, but in the restored colonial village of Williamsburg, in Virginia, there is an exact replica of Bourn mill (plate 41).

Most of the existing mills in America are small wooden smock mills, originally built in the eighteenth century. The development of the English and European smock mills described in chapter 8 was not finalised until the early nineteenth century, so the American mills were in fact copies of the types existing in England and Europe before that time. Not one example of this earlier type remains in England. In cap design particularly, the American windmills have much more in common with the French and Mediterranean mills (plates 85, 110) which still survive. It is as if no further development of windmill design took place in America after the millwrights built the first windmills there. Much later on, some mills incorporating fantails and patent shutters were built, probably by millwrights freshly arrived from Europe with the new ideas developed there.

The brick or stone tower mill with a wooden cap, so common in Britain, does not seem to have been at all popular in America. This may have been because there was a plentiful supply of timber with skilled carpenters and shipwrights to work it, but few masons and bricklayers, who were all otherwise engaged. The areas with most windmills were also those where much shipbuilding went on, and traditionally it was the shipwrights who built the early windmills.

The Dutch, who above all developed the windmill in Europe, were the earliest settlers of New York (then New Amsterdam), old drawings of which show many windmills. Part of the seal of New York is the sails of a windmill; a law, passed in 1680, forbade anyone to row across the river from New York to Brooklyn when the windmills were stopped and the cloths taken in because of high winds.

Later on, the Dutch devoted their millwrighting energies to building tidal mills and left the construction of windmills to the English.

Generally speaking, the mills in America have been rebuilt and extensively repaired, in some cases with more regard to looks than accuracy. These wooden smock mills have caps which do not resemble English types, though some are a little like the Lancashire boat-shaped cap (plate 77). They are winded by means of a tailpole from the back of the cap, usually on a waggon wheel at the base to take the weight. Others have hand winding wheels (plate 36). The structures are usually clad with shingles, not with weatherboarding as in the English mill. Shingle cladding was common to much wooden building in America at that time, so logically it was used for windmills. On the outside of some mills, iron rings were set into the cant posts, to which the sails could be anchored with chains when the wind was really strong.

PLATE 109A (opposite)
EASTHAM WINDMILL
MASSACHUSETTS
This shingle-covered mill has a conical cap with a dormer through which the windshaft emerges.

These are some of the existing windmills in the United States:

KENMARE WINDMILL, North Dakota. This was built by a Danish immigrant in 1887, so is in fact a late model. Maple was used for the gear wheels and the mill worked until the World War I era. The sails had a diameter of 36ft, which is not very big. The mill is recorded as being capable of 'turning out 200 sacks of grain [sic] a day'. Even if the record means flour, this is an enormous output. Of course, it does depend upon the size of the sacks!

EAST HAMPTON, Long Island, N.Y., had more windmills than anywhere else, and in the 1880s was a favourite place for artists. It has been said that 'the artists' easels and umbrellas were so thick that a farmer could hardly get to his own barnyard'.

HOOK MILL, East Hampton, owned by the village, stands on Memorial Green at the north end of Main Street. It is the third successive mill to be operated on the site. The original mill was built in 1796 by Nathaniel Doming.

The present structure began grinding in 1806 and was repaired in 1939 after lying idle for thirty years. It has a wooden windshaft through which pass the stocks of the double-sided common sails with hemlaths. The sails are weathered.

PANTIGO MILL, East Hampton, was built or perhaps rebuilt in 1771 for one Abram Gardiner. It has double-sided common sails with bars and hemlaths on long thin stocks mortised through the windshaft. The cap is round at its base, but has a little dormer hatch above the windshaft to allow access to the sweeps, which are of very light construction. The originals were probably much more rugged.

GARDINER MILL, East Hampton, was built in 1771. This is a smock mill with a conical cap which has gabled dormers back and front. The sails are double sided, with bars and hemlaths, and have a very pronounced twist or 'weather'. They are of much more solid construction than Pantigo mill and look pretty efficient.

NANTUCKET WINDMILL was built of old ships' timbers in 1764. Now it is locked and faces due west so can only be set going when the wind is in that quarter. The stocks are mortised through a slightly angled windshaft, and the common sails with leader boards have constant pitch. The cap, shaped like a house roof on a circular base with gable ends, turned on the shingle tower and was winded by means of a heavy tailpole at its lower end to take the weight. A little weather vane is mounted at the top of the back gable.

JAMESTOWN MILL, Jamestown, R.I. Built in 1787, it too has common sails, in this case single sided. There is a hand-winding wheel at the side of the cap.

EASTHAM WINDMILL, Cape Cod. This is Cape Cod's oldest windmill. It has a horizontal windshaft and a conical cap from the back of which protrudes a long beam, described as a brake handle. There appears to be no tailpole or other means of turning the conical cap on its shingle-clad tower. The double-sided common sails have constant pitch. One cannot help feeling that this reconstructed mill differs somewhat from the original.

One of the oldest windmills on Cape Cod stood just across Lewis Bay, in Yarmouth Town, until it was moved to Dearborn, in Michigan, by the late Henry Ford.

BRIDGEHAMPTON MILL, Long Island. This looks much more like an English mill than any other as it has a fantail on a staging and an ogee-shaped cap. It has double-sided common sails with hemlaths.

PORTSMOUTH MILL, Long Island. This is a total reconstruction based on an old mill originally built in 1790. It has common sails with leader boards, and is winded by means of a wheel. An endless chain passes over the wheel to ground level, where the miller can reach it and turn the wheel so that the cap turns on its curb. The cap is hemispherical with access doors fore and aft to sails and winding wheel.

WINDMILL AT WATERMILL, Long Island, has a conical cap with gabled dormers out of which pass the windshaft and the massive tailpole. It has double-sided common sails.

CHATHAM MILL, Massachusetts. This smock mill, built in 1797, with a cap shaped somewhat like a house roof, was winded by means of a tailpole and base wheel. It is built on a man-made mound to catch just that little bit more wind and to clear nearby trees.

PLATE 109B
CHATHAM WINDMILL
MASSACHUSETTS

A shingled tower mill with constant pitch sails. The mill has a gabled cap, and is winded by means of a tailpole from the back of the cap with a wheel to support it at ground level.

PLATE 110 (opposite page)
MILL NEAR LA MANGA, SPAIN

This tower mill, with its scalloped boards and massive tailpole, pumped up water and then moved it into irrigation channels by means of a scoop wheel, which is just visible behind the tower. The framework of the wooden poles and wire rigging is in good repair, and one can see exactly how the stresses of the sails were taken. The triangular cloth sails were set on the poles and rigging as the sail of a boat is set on mast and boom.

PLATE 111
WINDMILL ON MAJORCA
BALEARIC ISLANDS
This mill was advertising a restaurant, but had once been used for grinding. Cloth sails are set on sail bars attached to the sail backs. The poll end appears to have an iron circle behind it to which the sail backs are bolted. Photograph taken at 1/15 second to catch the movement of the sails.

There is a windmill converted into a residence on the campus of Southampton College, Southampton, Long Island. It was built around 1800, and is like Hook mill in basic type.

In the Golden Gate Park in San Francisco there are two smock mills built comparatively recently to pump water for the houses and works there. These mills have fantails and common sails.

Mediterranean and Iberian Mills

Anyone who has travelled in Spain or Portugal and visited Mediterranean islands will be familiar with the sight of hundreds of little tower mills, many of them working. Coming in to land at Palma airport, on Majorca, one sees below a patchwork of small green fields, each with its little modern windmill on a tower. On Crete there are thousands of these little mills of ancient type, with four to six triangular cloth sails spread on a wood-and-rope or wire frame, winded by a vane. Both these types of mill – on Crete and on Majorca – are lifting water from just below the surface through simple pumps to irrigate the fields.

In mainland Spain and Portugal and elsewhere one sees bigger tower mills, also with triangular sails on a framework, usually for grinding grain. Also on Majorca and in mainland Spain one finds tower mills, for pumping or grinding, with much bigger cloth sails set on a backing of sail bars attached to a framework, and with rigging between sail ends and between a central pole which projects forwards to take the backward pull of the sails. The Spanish mill has a scoop wheel. The Majorcan mill (plate 111) was being used purely to draw tourists to a restaurant, but at least it was turning.

The advantage of the triangular jib sail is that it can be reefed and reduced to a tiny area when the wind blows up. Some Portuguese mills carry whistles made of clay, attached to the rigging, which sing out when the wind catches them at a certain angle, warning the miller working nearby in his fields that the wind has changed and the mill needs winding.

The machinery of Mediterranean and Iberian grinding mills does not differ in basic principle from that of mills anywhere else, but the variations are infinite and there is much comparative study to be done.

Portugal has a very active windmill society; the windmills there form an integral part of rural life, although more and more are being replaced by modern machinery. In the other Mediterranean countries, it will be left until only a few mills remain in out of the way places before any concerted efforts at preservation are made.

Mills in Germany and Eastern Europe

All the countries of Europe have some windmills, but unfortunately I do not have room in this book to describe them all. In Germany, particularly in Schleswig Holstein, that area south of Denmark and north of Holland, there are perhaps fifty windmills in various states of preservation and dilapidation. North-west of Hamburg, north of the river estuary, there are twenty mills, many preserved, which are well worth a visit. These German mills have many features in common with Dutch and Danish mills. There are a few mills in Saxony and in other parts of Germany.

Russia, Poland and the Balkan countries also have some windmills, of course, and many thousands of watermills. Romania in particular has many mills, mostly post mills with six canvas sails, but also some post mills with Mediterranean-type jib sails, and some 'Dutch type' mills with movable caps. A mill hunting holiday in this country could be fascinating, especially as the language difficulties would be enormous!

15

Millwrighting

The modern millwright is very seldom involved in building a mill from scratch, and for that reason alone his work is very difficult. He has to find ways to replace rotten and broken timbers without dismantling entire sections of the mill. Because mills were put together in a very definite order, timbers are locked in place by their surrounding pieces and, although they can be cut out, it is not always so easy to get replacements back in. Very often he cannot make mortise and tenon joints but must make do by strengthening overlapping joints with steel plates and brackets.

Millwrighting is a dangerous job (plate 95) and there are few safeguards. The millwright must climb sails, and get up on to fantail stages and cap ridges, often pretty rotten, without being able to hook himself on with safety harness. He must have the attributes of a steeplejack – perfect balance and a head for heights. He must also be prepared to work alone for a lot of the time. At least in the old days, when a mill needed repair, the millwright was never alone and had such help as he needed to hand. He is not a member of a union, and has no shop steward to prevent him doing the job of a painter, bricklayer, carpenter, joiner, metal worker, blacksmith, architect, or surveyor, in addition to being a craftsman and a perfectionist.

The millwright in plate 112 is putting a taper on three sides of a 42ft-long balk of pitch pine, to make it into a mill stock. This could be done on a bench with a circular saw or with a chain saw, but few sawmills have the capacity to do this and the chain saw must be very carefully handled, as it can cut off line so quickly and spoil the whole job. The adze is no different from the ancient tool which was one of the earliest developed as man emerged from the Stone Age. This adze is made of mild tempered steel and is so sharp that, should it slip or skid, it could do serious damage to the worker's leg. So, having ruled the timber, the millwright chips away carefully till he has reached the point where a smoothing plane finishes off the job. This particular balk of wood, 42ft by 1ft square, shows clearly, where it has been adzed, that it was not cut from a very big tree. The dark heart wood forms about 75 per cent of the balk at one end, but tapers off to about 25 per cent at the other, which is mainly sap wood. The heart wood does not run absolutely true and straight right up the balk, nevertheless it is the best timber obtainable.

In the old days, for this kind of job, very big trees were taken out of virgin forests, where they had been growing surrounded by dense timber and had reached straight upwards to the light above them. The tree from which the above balk was cut probably in open woodland, with light coming through the trees; as the seasons and years passed, it slowly twisted towards the south and the sunshine.

Seasoned timber of any kind is extremely hard to come by nowadays – that is, timber properly

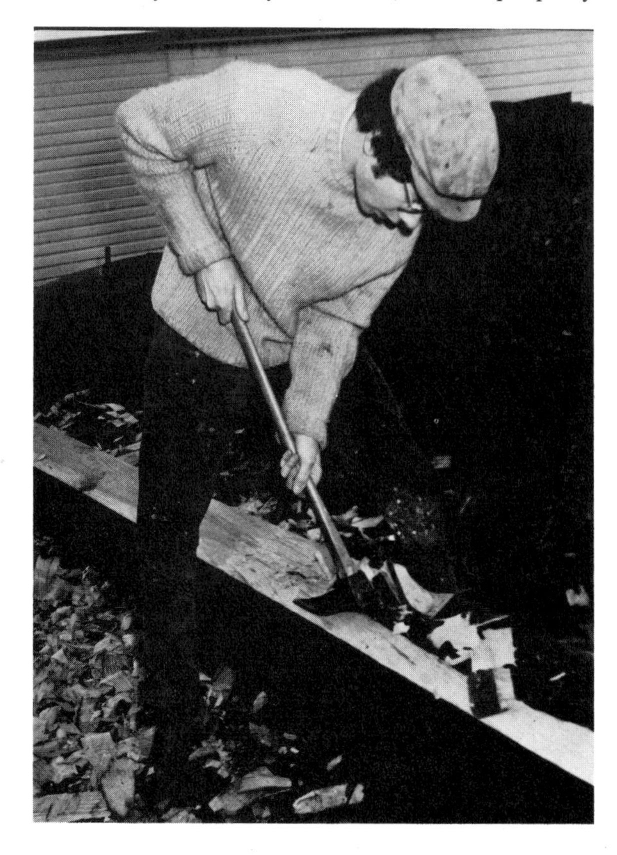

PLATE 112
TAPERING A SAILSTOCK
The millwright is using an adze to taper this pitch-pine balk.

seasoned by being stacked out of doors for years before it is used. Modern timber, kiln dried or seasoned for too short a period, comes to the millwright still containing too high a percentage of water; after it has been worked and set up, the seasoning continues in the wind and rain and sunshine, and all kinds of warping and splitting may occur. It is a fact of course, that, however well seasoned wood may be, it still goes on drying out for years and years, and this is very evident in the ancient French mill, 'Michel' (plate 88) where shakes have been bound together with iron clamps to strengthen the timbers.

I was recently asked by a man who prides himself on being modern in all things if a windmill could not be built much more efficiently and beautifully using metal alloys and synthetics, rather than timber and castings. 'How it would gleam,' he said, 'and how smoothly it would run!' So it might, but for how long? Would modern alloys and fibre glass stand up to centuries of use and weather, violent stresses and changes of temperature, as have wood, brick and cast iron and steel? Surely it is the inherent natural qualities of wood which makes it the best possible material. The millwright could choose each different wood for its quality suitable for his purpose; for instance, pitch pine for sails, because it is straightforward and resistant to decay, and oak for the main post, needing a piece about 2ft 6in square, 18–20ft high and weighing about 1½ tons. The cross trees and quarter bars were also made of oak. Some Sussex post mills had four-piece pitch-pine main posts, tapered, with hoops driven down round them from the top to bind them together. The crown tree could be made the same way; Clayton post mill has a pitch-pine crown tree. Normally the crown tree, weighing about a ton, and all the main timbers, are of oak. Oaks for mills were carefully selected and millwrights preferred hedgerow trees growing on heavy stony soil. Trees were usually barked before being felled, and the bark used for tanning. Spring barked trees felled in the autumn provided tough wood which was least susceptible to woodworm. This was because through the summer the foliage took up all the sap and by autumn the sapwood would be nearly as hard as the heartwood.

Nowadays oak is cut when it is about sixty years old, but the old mills were made of 100-year-old wood. In any case, oak trees are protected and only those which get in the way of modern developments may be cut down so English oak of any size is scarce indeed.

Pitch pine was also used in many mills, and for modern repair work is almost invariably the best timber available. Now, it is getting more and more difficult to find good pitch pine, and Douglas fir is the next best thing. Not reckoned to be a long-lived wood, good Douglas fir is still better than bad pitch pine. Sweet chestnut was sometimes used, and cap ribs were made from elm, ash and poplar.

Tar is the best preservative for mills, and present-day millwrights paint their timbers with preservatives, such as Cuprinol, as they work, in an attempt to get them thoroughly impregnated. White paint looks best of all, but as about 3 cwt is needed to paint a big mill properly it can be a very expensive job, and unless paintwork is kept up it cracks and exposes underlying timber, and creates little ledges where water can collect and seep into the timber. Modern paint is purely decorative and has no preservative properties at all. It puts a waterproof plastic skin on a surface and consequently any moisture which gets behind it cannot dry out. In a structure such as a windmill it is impossible to prevent moisture getting in between timbers. Post mills on the whole last longer than tower and smock mills, but some enormous tower mills were erected. Southtown mill at Gorleston cost £10,000, an enormous amount of money in 1812, and soared up to 122ft, many feet higher than Sutton (plate 74). Its sails spanned 84ft. This kind of building needed very good foundations indeed to avoid eventual subsidence.

One of the most expensive aspects of mill repair is the re-making of metal parts. The great cast-iron windshaft which could be made for £38 in 1880 now costs about £1,000, if you can find someone to cast it. Of course, much use is made of spare bits and pieces from mills that have been dismantled, but where new pieces are required they have to be custom built. Connecting rods, spindles, levers, screws, etc, can be made by a competent smith, very often the millwright himself, from suitably sized blanks, but parts which have to be cast require a different technique. The millwright must first make a perfect wooden pattern, and this requires absolutely accurate measurement in the first place. He then must combine the skill of carpenter and wood carver to produce the model. This he will make in soft wood and finish by rubbing it down to glass smoothness. It is to be used as a pattern from which to make a mould, and each and every slight imperfection will be reproduced in the finished casting.

For cogs, apple wood, seasoned for nine years, is the best, because it will shrink no further and will stand up to the continual wear of meshing gears. Hornbeam and beech are also used. Each wheel has to have exact mortises cut to take the cogs; the marking out and working of these mortises and cogs takes accurate craftsmanship of a high order. One loose-fitting cog can cause excess strain to be put on the other cogs on the wheel.

Putting in New Stocks

The new stock, some 40ft long, has been cut: tapered at each end on three sides, with the fourth side – the front of the stock – left flat, and the centre carefully cut slightly smaller than the inside measurement of the canister.

The millwright then fixes a four-to-one tackle to the end of the canister, which in some cases has a convenient knob (plate 12), to which he can attach the top pulley block. Otherwise he will have to lash it very firmly to the canister itself in such a way that it will not interfere with the stock as it is hauled up. The lower block is then attached by a sling about 4 or 5ft down the stock, which is laid out on the ground in front of the mill, at right angles to it. The men haul on the tackle and the stock swings up so that the top end passes into the canister. The slinging is all important. A wrongly slung stock will not go up neatly. When the stock is up to the point where its lower end rests on the ground (or on the outside stage if there is one), the sling is re-set lower down and the stock pulled further up the canister. If the stock does not reach the ground after the first pull up, then the millwright drills a hole through it and puts in a bolt with a wooden stop bar on one end so that the stock cannot drop back down when the sling is taken off.

FIGURE 31 Putting in stocks.

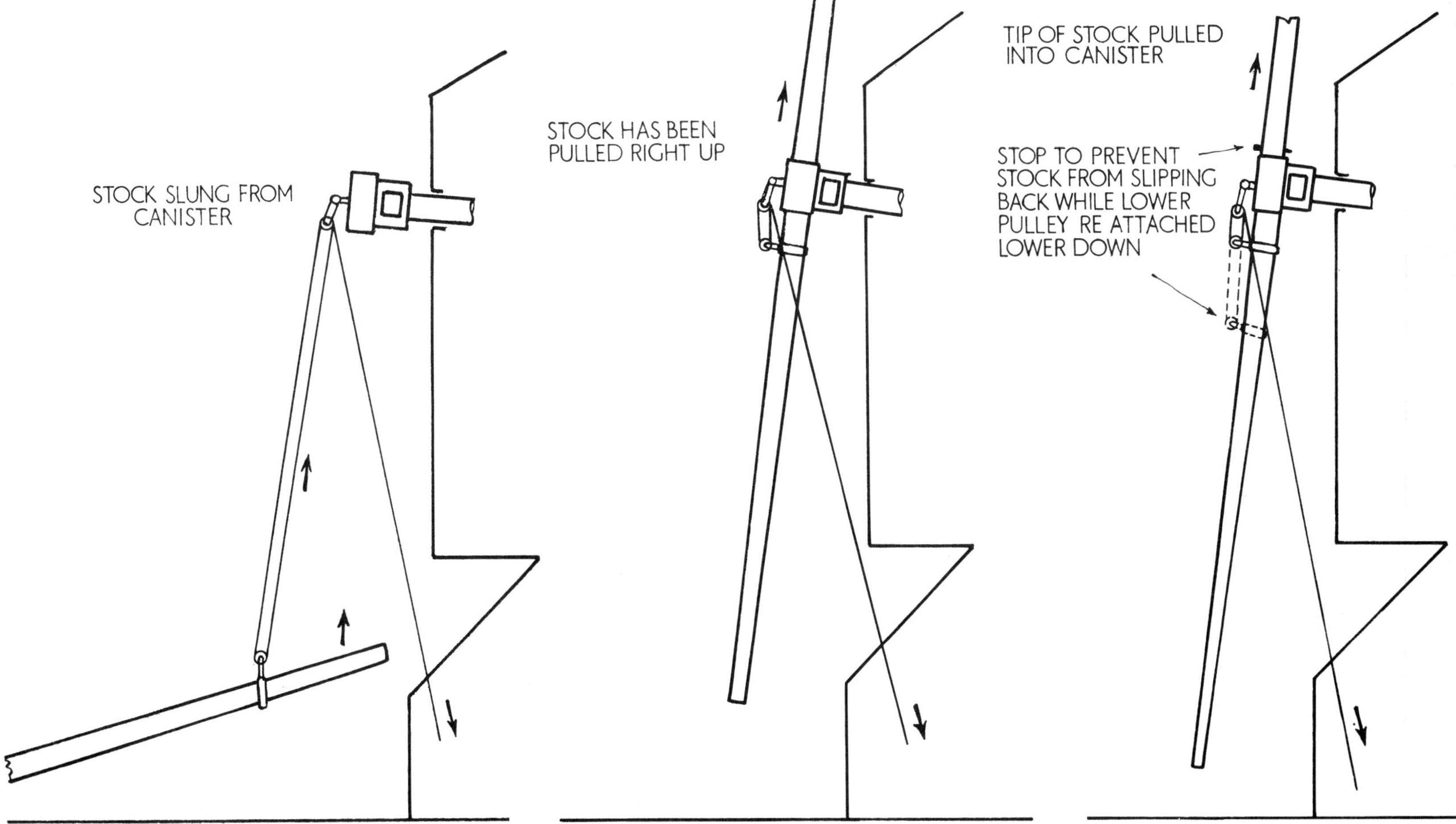

PLATE 113
MOUNTNESSING MILL, ESSEX
A new whip being pulled up into position. Note the ropes tied to the end of the upper stock and the left-hand sail, by which the sails can be pulled round. The lower pulley is attached high on the whip, and the upper pulley to the poll end.

After the second pull up, the stock should be most of the way home, and at this stage the mill is turned by pulling down on one end of the other pair of sails or stocks, to which a rope has been attached (plate 113). When the new stock is at twenty-five past eleven, the whole contraption is shaken and jogged until the stock has slid into position, where it can be secured with wedges. If the stock centre has been cut too big, all kinds of problems ensue, for it will jam and will not go into position, but the millwright must resist a temptation to cut it too small or, even when wedged, the stock will rock in the canister and cause undue stresses as the mill turns.

Should the stock be so long that the tip cannot be got inside the canister when the pulleys are attached centrally, then the top pulley must be attached to the other stock and the mill turned so that the point of attachment is higher than the canister. It is not quite so easy to persuade the new stock into the canister, but guided by ropes it can be done.

The sails must then be added to the stocks; these too are pulled up one by one using tackle attached to the canister or to the stocks themselves.

To make and fit new stocks and whips, make up the sails with bars, backstays, etc, and fit them, takes about four months. It also took about four months for two men to re-make the cap in plate 114. The timbers were hauled up by tackle on a long pole, which the millwrights fixed in the top of the mill, jammed right down below the brake wheel. This pole stood up high above the curb so that the timbers could be hauled up and dropped directly into position.

Apart from blocks and tackles, a cradle has many uses. Secured firmly, it can be raised and lowered by the men in it, and from it they can get at otherwise inaccessible parts of the outside of the mill without using ladders.

To restore a smock mill which is nearing the state where it will break up if nothing is done, but while it still has its original machinery, will take about two years and cost upwards of £12,000.

(a)

(b)

(c)

PLATE 114
WHITE RODING MILL, ESSEX
Building a new cap frame, a job which took four months: (a) One of the sheers is being hauled up. Note the pole jammed under the brake wheel to act as a hoist. The falls of the hoisting tackle can be seen going down diagonally each side of the mill, to where the men at the bottom were pulling; (b) The cap frame is built and the petticoat is on. Only the 18 sw9 aluminium sheeting cover still has to be put on; (c) The finished job. Note the sheers projecting forward under the weather beam.

PLATE 115 (overleaf)
SANDWICH MILL, KENT
The cradle is hung from the fantail stage so that the millwrights can get up and remove the hanging fantail, which could not have been reached by ladder.

The Future

There are only four or five millwrights working today, and they have full work books at the moment, but what of the future? Not only do materials get more and more expensive, but so do living costs, and the craftsman millwright deserves a good living. Should the trade die out those mills which have already been restored will eventually be looked after by builders who do not know the finer points of millwrighting. Unfortunately, this is already happening to some extent.

It is often hard to justify an expenditure of several thousands of pounds by a town or parish council when there are so many other more pressing needs. While voluntary efforts can raise some money, they can rarely raise enough, and a millwright cannot just be on call to do a few pounds' worth of work at a time – although his concern for windmills does in fact often lead him to do this to prevent further deterioration. Yet it is surely a matter of relative priorities. Everything costs much more – public lavatories as much as windmill repairs – but once windmills are gone they are gone for good. It is for the future we preserve. We look back with gratitude, or enough of us do, to those who have preserved things in the past. The *Cutty Sark*, for instance, very nearly got broken up for scrap. Brighton Pavilion was bought by Brighton Corporation from Queen Victoria in 1850 when it would otherwise have been demolished. There are hundreds of examples all over Europe.

Thankfully, craftsmen do persist – smiths, thatchers, millwrights, boatbuilders and others, many of them young men. In our time of mass production and machine technology we do not value them highly enough. They should be among the élite of our age because they alone cherish and perpetuate man's ability to create with his hands and with hand tools from natural materials, the things he needs and it is a fact that things created out of man's necessity in this way are usually beautiful. There is harmony of shape and line and mass and colour, which is all the more satisfying because it has not been self-consciously sought after. The millwright is not trying to get a message across. He is trying to create something which is strong and will do its job for many years, because it is well made of the right materials, and he instinctively works within the natural laws of balance and harmony. Almost accidentally, he therefore creates beauty. The wind and the weather, the sky and the sunlight accept and enhance that beauty; the least that we can do is to try to preserve it.

The Old Mill at Night

Broken you stand.
Your sails are crooked skeletons
against a gibbet sky.
Your innards stink of rats and rotten grain
and cobwebs choke the filthy window pane.
Millstones lie frozen and the gap-toothed wheels are still
for no one any longer tends the mill.

Yet in the blackest night perhaps you shrug
your dirty shoulders, stretch and standing tall
you lift your face towards the rising wind.
White floury ghosts of millers hump the grain
and thumb the meal while down and down again
your sails go scything through the screaming sky.

Till dawn comes ragged and for one more day you die.

SUZANNE BEEDELL

Appendix 1

Where to find windmills

No up-to-date absolutely comprehensive list of mills exists, although SPAB publishes an up-to-date list of those open to the public. The status of windmills changes all the time and my own list, unavoidably, will be slightly out of date by the time this book is published. Some counties are very short on mills; this may be because little work has been done on recording or measuring there, not because the mills do not exist.

In the list below, all the mills marked with an asterisk are on SPAB's list and are worth a visit. Other mills which I know to be in reasonable order and worth a special visit are marked with a cross, although they may not necessarily be open. Question-marked mills may or may not be of great interest; I have been unable to find out their present status and you must take a chance on these. Mills marked with an 'M' have been mentioned in this book. Many mills are listed here as derelict (D); some of these may be preserved in this condition, rather than being allowed to fall down altogether. Even though not marked with an asterisk or a cross in this list, derelict mills may have interesting features.

One important word of warning: if a mill stands in private grounds, and has no notice stating that it is open to the public, DO NOT TRESPASS. Treat a mill which is private, or is somebody's home, exactly as you would any other private property – with respect.

KEY

*	SPAB-listed and worth a visit.
M	Mentioned in book.
D	Derelict.
+	Worth a visit.
C	Converted to a house.
?	Condition not known.

ANGLESEY

?	Landeusant tower mill
D	Treaddur Bay tower mill

BEDFORDSHIRE

C	Duloe tower mill, Eaton Socon
D	Houghton Conquest tower mill
C	Sharnbrook tower mill
*	Stevington post mill
D	Thurleigh tower mill
D	Totternhoe tower mill
D	Upper Dean tower mill

BUCKINGHAMSHIRE

+	Bradwell tower mill
*	Brill post mill
C	Cholesbury tower mill
+	Coleshill tower mill
?	Fulmer tower mill
C	Ibstone smock mill
M*	Ivinghoe, Pitstone post mill
M+	Lacey Green smock mill
D	Quainton tower mill
C	Wendover tower mill

CAMBRIDGESHIRE

M*	Bourn post mill
+	Burwell tower mill
D	Cambridge, Chesterton smock mill
*	Great Chishill post mill
M	Fulbourn smock mill
D	Gamlingay smock mill
D	Guilden Morden tower mill
D	Haddenham tower mill
+	Histon smock mill
+	Madingley post mill
M+	Over tower mill
D	Soham tower mill and two smock mills
D	Six Mile Bottom post mill
D	Steeple Morden smock mill
+	Swaffham Prior tower and smock mill
M+	West Wratting smock mill
*	Wicken Fen smock pumping mill
+	Wicken smock mill
C	Little Wilbraham tower mill
D	Willingham smock mill

CHESHIRE

M+	Bidston tower mill
D	Capenhurst tower mill
D	Gayton tower mill
C	Saughall tower mill
C	Willaston tower mill

DERBYSHIRE

*	Dale Abbey, Cat & Fiddle post mill
+	Heage tower mill

DORSET

D	Portland, Easton tower mill

DURHAM

?	Greatham tower mill
+	Sunderland, Fulwell tower mill

ESSEX

D	Ashdon post mill
+	Aythorpe Roding post mill
C	Great Bardfield, Gibraltar tower mill
*	Bocking post mill
+	Clavering, two tower mills
C	Great Dunmow tower mill
+	Finchingfield post mill
+	Fryerning post mill
M+	Mountnessing post mill
D	South Ockendon smock mill
D	Orsett, Baker Street smock mill
D	Ramsey post mill
+	Rayleigh tower mill
*	Stansted tower mill
*	Stock tower mill
MC	Terling smock mill
+	Thaxted tower mill
C	Tiptree tower mill
D	Toppesfield, Gainsford End tower mill
*	Upminster smock mill
+	White Roding tower mill

GLOUCESTERSHIRE

?	Falfield tower mill
?	Frampton Cotterel tower mill
?	Warmley tower mill

HAMPSHIRE

+	Chalton tower mill

HERTFORDSHIRE

+	Arkely tower mill
?	Brent Pelham smock mill
+	Cromer post mill
D	Little Hadham smock mill
?	Tring, Goldfield tower mill

HUNTINGDONSHIRE

+	Barnack, Peterborough tower mill
*	Great Gransden post mill
C	Hemingford Grey tower mill
+	Whittlesey tower mill
D	Yaxley tower mill

ISLE OF WIGHT

M*	Bembridge tower mill

KENT

D	Benenden smock mill
?	Bidborough tower mill
C	Canterbury, St Martin's Hill tower mill
+	Charing smock mill
M*	Chillenden post mill
+	Chislet smock mill
M+	Cranbrook smock mill
M+	Eastry smock mill
M+	Herne smock mill
D	Keston post mill
*	West Kingsdown smock mill
D	Kingston, Reed tower mill

M* Margate, Drapers smock mill
* Meopham Green smock mill
D Northbourne smock mill
+ Rolvenden post mill
CM St Margaret's Bay smock mill
M+ Sandwich smock mill
D Sarre smock mill
D Shorne post mill
D Stanford tower mill
M+ Stelling Minnis smock mill
C Whitstable, Borstal Hill smock mill
C+ Willesborough smock mill
+ Wittersham, Stocks post mill
D Woodchurch smock mill

LANCASHIRE

D Ainsdale tower mill
C Bretherton tower mill
D Clifton tower mill
D Great Crosby tower mill
+ Haigh tower mill
? Lytham St Annes tower mill
+ Little Marton tower mill
D Pilling tower mill
D Staining tower mill
M+ Thornton Cleveleys tower mill

LEICESTERSHIRE

D Arnesby tower mill
D Gilmorton tower mill
* Kibworth Harcourt post mill
D Waltham-on-the-Wolds tower mill
? Wymondham tower mill

LINCOLNSHIRE

* Alford tower mill
D Belton, Westgate tower mill
+ Burgh-le-Marsh, two tower mills
D Dyke smock mill
D Heapham tower mill
* Heckington tower mill
+ Horncastle tower mill
D Kirton-in-Lindsey tower mill
D Lutton, Sutton Bank, tower mill
+ Sibsey, Trader tower mill
M+ Boston, Skirbeck, Maud Foster tower mill
D Sturton-by-Stow tower mill
+ Waltham tower mill
* Wrawby post mill

There are, in Lincolnshire, the remains of many other tower mills, usually just an empty tower with little or nothing inside. At one time gipsy scrap men went round buying up all the old metal parts of these mills, which accounts for the state of many of them.

INNER LONDON

+ Brixton tower mill
+ Wandsworth Common smock mill
+ Wimbledon Common smock mill

NORFOLK

D Blakeney tower mill
* Billingford tower mill
D Old Buckenham tower mill
C+ Burnham Overy Staithe tower mill
D Carbrooke tower mill
+ Caston tower mill
MC+ Cley-next-the-Sea tower mill
+ Denver tower mill
D East Dereham tower mill
D East Harling tower mill
+ Garboldisham post mill
D Hempnall tower mill
C+ Hickling tower mill
C Honing tower mill
D Horsford tower mill
C+ Paston tower mill
C Ringstead tower mill
D East Runton tower mill
M+D Sutton tower mill
D West Winch tower mill
MC Weybourne tower mill
D Wicklewood tower mill

NORFOLK BROADS AREA DRAINAGE MILLS

D Acle hollow post mill
D+ Ashtree Farm tower mill
M+* Berney Arms tower mill
D Brograve tower mill
MD Chedgrave Detached tower mill
C Clippesby tower mill
D Haddiscoe, Toft Monks Detached tower mill
D Halvergate Marshes, four tower mills
D Heigham Holmes tower mill
D Hickling, Stubb tower mill
D Horning Ferry open trestle mill
C Horning Ferry smock mill
M* Horsey Mere tower mill
MD How Hill open trestle mill
D How Hill tower mill
? Hunsett tower mill
D Langley, two tower mills
C Langley Detached tower mill
D Lockgate, Breydon North Wall, tower mill
D Ludham Bridge North Side tower mill
C Martham Level tower mill
D Mautby Marsh tower mill
D Oby tower mill
C Potter Heigham, Highs tower mill
D Seven Mile House, three tower mills
D Runham, three tower mills
MD Runham, Six Mile House tower mill
D Runham, Key's tower mill
D West Somerton tower mill
+ Stracey Arms tower mill
M+ Thurne Mouth, two tower mills
C Upton tower mill
D Waxham, Lambrigg tower mill
D Womack Water tower mill

NORTHUMBERLAND

+ Woodhorn tower mill

NOTTINGHAMSHIRE

D Gringley-on-the-Hill tower mill
M* North Leverton tower mill

OXFORDSHIRE

+ Bloxham Grove, Bodicote post mill
D Great Haseley tower mill
D Wheatley tower mill

RUTLAND

+ Morcott tower mill
+ Whissendine tower mill

SHROPSHIRE

? Much Wenlock tower mill
D Vennington tower mill
D Weston tower mill

SOMERSET

* Ashton tower mill
? Brockely Combe tower mill
? Clevedon, Kenn tower mill
? Felton Common tower mill
* High Ham tower mill
? Hutton tower mill
? Locking tower mill
? Portishead tower mill
? Walton, near Street, tower mill
? Watchfield tower mill
? Weston-super-Mare, Uphill tower mill
? Worley tower mill

SUFFOLK

D Bardwell tower mill
D Belton, two tower mills
D Buxhall tower mill
+ Dalham smock mill
M* Drinkstone post mill and smock mill
* Framsden post mill
D Friston post mill
D Fritton tower mill
* Herringfleet smock mill
M* Holton St Peter post mill
+ Minsmere smock mill (in nature reserve)
M* Pakenham tower mill
D Reydon Quay tower mill
+ St Olaves trestle smock mill
M* Saxtead Green post mill
D Stanton post mill
D Syleham post mill
D Thelnetham tower mill
D Thorpeness post mill
+ Great Thurlow smock mill
D Great Welnetham tower mill
* Woodbridge, Buttrums tower mill

SURREY

C Hurtwood Common, Ewhurst, tower mill
+ Lowfield Heath post mill
* Outwood Common post mill
* Reigate post mill
+ Reigate, Wray Common tower mill
* Shirley tower mill
D Tadworth post mill

SUSSEX

M+ Argos Hill, near Mayfield, post mill
? Arundel tower mill
+ Barnham tower mill
C Battle smock mill

+	West Blatchington smock mill
+	Chailey smock mill
C	West Chiltington smock mill
M+	Clayton, 'Jack', tower mill
M+	Clayton, 'Jill', post mill
M+	Cross-in-Hand post mill
C	Crowborough tower mill
+	Ditchling, Oldland post mill
D	Earnley smock mill
M+	Halnaker tower mill
M+	Icklesham, Hogg Hill post mill
D	Nutbourne tower mill
*	Nutley post mill
D	Nyetimber tower mill
C+	Patcham tower mill
M*	Polegate tower mill
M+	Punnetts Town, Blackdown smock mill
+	Rottingdean smock mill
M	Rye smock mill
*	High Salvington post mill
+	Selsey, Medmerry tower mill
M*	Shipley smock mill
+	Stone Cross, Pevensey, tower mill
MC	Washington smock mill
D	Windmill Hill, near Herstmonceux, post mill
+	Winchelsea post mill
D	East Wittering, Selsey, tower mill

WARWICKSHIRE

?	Boarstall Common tower mill
?	Burton Dasset tower mill
*	Chesterton tower mill
+	Compton Wynyates tower mill
++	Danzey Green post mill (now standing in Rural Life Park, Bromsgrove)
+	Napton tower mill
*	Norton Lindsey tower mill
?	Thurlaston tower mill
D	Whitnash tower mill

WILTSHIRE

+	Wilton tower mill

YORKSHIRE

+	Elvington tower mill
+	Holgate, York, tower mill
*	Skidby tower mill

Appendix 2
Windmill societies and museums

BRITAIN

Society for the Protection of Ancient Buildings
Wind and Watermill Section
55 Great Ormond Street
London WC1N 3JA

DENMARK

National Museum, Molinology Laboratory
Brede Voerk
DK 2800
Lyngby
Denmark

FRANCE

Association Française des Amis des Moulins
Musée National des Arts et Traditions Populaires
6 Route de Madrid (Bois de Boulogne)
75 Paris 16e

HOLLAND

Association for the Preservation of Windmills in the Netherlands
De Hollandsche Molen
Amsterdam C
9 Reguliersgracht

PORTUGAL

Associacão Portuguesa de Amigos dos Moinhos
Mueseo de Arte Populaire
Lisbon
Portugal

ROMANIA

Brukenthal Museum
Sibiu Hermanstadt
Romania

USA

Society for the Preservation of New England Antiquities
Harrison Gray Otis House
141 Cambridge Street
Boston, Mass
USA

Glossary

air brakes longitudinal boards in outer end of leading edge of sail, actuated by shutter mechanism to open and break up air flow in heavy gusts.
back stays supporting bars across the back of the sail.
backwinded mill caught by wind when sails are not facing the wind.
beard decorated board behind the canister on cap of Dutch mills.
bedstone the lower of a pair of millstones, which remains stationary.
beehive cap Kentish name for a domed cap.
bell crank lever part of spider which links striking rod to uplongs.
bills wedge-shaped pieces of metal used for chipping (dressing) millstones.
birdbeaks notches at the junction of quarter bar and cross trees.
bist rough cushion to ease the stone dresser.
bolter type of flour dresser machine with woollen sleeve.
brake wooden or iron brake shoe encircling brake wheel.
brake wheel cog wheel mounted on windshaft which drives the wallower and round the rim of which the brake contracts to stop the mill.
bran partly ground husk of grain.
brayer lever beam on which the bridge tree rests.
breast the lower part of the front of the buck of a post mill which protects the trestle from the weather. The part of the buck in front of the post.
breast beam main lateral beam beneath the windshaft.
bridge metal bar cemented into eye of runner stone to act as bearing for the top of the spindle.
bridge tree lever beam which carries the lower end of the spindle and thus bears the weight of the runner stone.
buck the whole body of a post mill above the trestle which revolves as the mill is winded.
canister large cast-iron double socket on the end of the windshaft through which the stocks pass.
cant posts main corner beams of sections of a smock mill.
cap movable top section of tower or smock mill.
cap circle lower bearing surface of cap which rests on the curb.
centring wheels wheels attached to cap frame which centre the cap in relation to the tower.
chain wheel wheel turned by means of an endless chain. For winding mill or for working striking gear.
cloth sails sails with a wooden framework covered by cloth.
common sails cloth sails.
common toll toll exacted as payment for grinding corn.
constant pitch sails sails with bars set at identical angles to the whip from inner to outer end.
cracking cutting the fine grooves (drills, feathering or stitching) along the lands of a millstone.
cross trees large horizontal beams, the ends of which rest on masonry or brick piers at the base of the trestle on a post mill, and which carry the weight of the whole structure via the quarter bars.
crotch Y-shaped attachment to the quant which slots over the bridge into the mace on overdriven stones.
crown tree main beam across the body of the buck upon which pivots the top of the post carrying the whole buck with it.
crown wheel horizontal gear wheel engaging with the vertical gear wheel.
damsel contraption above the bridge on underdriven stones which causes the shoe to wobble, shaking grain down into the eye.
edge mill mill in which the stones run on their edges.
eye hole in the centre of the runner stone through which grain passes into the middle of the two stones.
fan staging the wooden supports and platform of the fan mechanism at the top of a tower or smock mill.
fantail small secondary windmill geared to turn the cap to face the wind.
fantail carriage carriage running on wheels on the ground carrying a fantail, and attached to the tailpole and/or ladder of a post mill.
feathering *see* **cracking**
feed shoe guides grain from hopper into eye of stone.
finial pointed top of an ogee cap, with a knob on it.
flour dresser machine for separating flour from the rest of the meal.
fly tackle another name for a fantail.
fork iron part of the spider mechanism which joins levers to the shutter bars.
French burrs prized millstones from France made of small blocks of fresh-water quartz fitted together.
furrow or furrow strips low part of pattern on the surfaces of millstones.
governor automatic device which adjusts the distance between the stones as the sails turn faster or slower.
grain hopper hopper above the vat which holds the grain to be milled.
great spur wheel mounted near the bottom of the upright shaft, it meshes with the stone nuts to drive the millstones. Also provides drive to other subsidiary machinery.
head upper front part of a post mill.
head and tail mill post mill with two sets of millstones, one geared from each end of windshaft.
heel inner end of sail.
hemlaths pieces of wood running longitudinally along the edges of sails to hold the bars firmly.
hollow post mill a post mill in which the drive passes down through the centre of the post to gearing at the bottom.
iron cross type of poll end common in some parts of England.
iron gudgeons part of the bearing at the junction of post and crown tree.
jib sails sails with leader boards which pass air *behind* main section of sail to provide 'lift' or suction, as on aircraft wings or dinghy sails. Or, triangular cloth sails on Mediterranean-type windmills.
keep irons iron bars so fitted to the cap as to hold it down to the tower should the mill be backwinded. The iron bar holding down the tail of the windshaft.
lands high parts of pattern on the surfaces of millstones.
lantern pinions gear wheel consisting of staves set between two discs.
leader boards longitudinal boards on the front or leading edge of a sail.
mace jaws at top of spindle which slot over the bridge thus providing the drive to the runner stone.
main post upright post on which a post mill revolves.
main shaft vertical shaft from wallower to spur wheel.
manyheight stepped wedge for use as fulcrum with a crowbar.
middlings intermediate product from flour dressing.
mill bill *see* **bills**
milling soke manorial law governing ownership, building and usage of mills.
millstones the pair of stones which grind the grain.
mortise wheel iron wheel with wooden cogs mortised into it.
neck bearing front bearing which supports the windshaft.
ogee cap domed cap with reverse curve at the top and a more or less pointed finial culminating in a knob.
open stones coarse uncut millstones.
overdrift mill with runner stone driven from above.
paint staff *see* **wood proof**
paltrok mill small Dutch sawmill which turns as a whole on a curb on a low wall.
patent sails Shuttered sails linked through a spider to an automatic opening and shutting mechanism.
pepper pot high domed cap with a flat top instead of a finial.
petticoat vertical boarding on cap to protect junction of cap and tower against the weather.
pintle part of the bearing at the junction of post and crown tree.
pit wheel vertical bevel wheel on same axle as scoop wheel which takes the drive from the crown wheel at the bottom of the vertical shaft.
polder mill Dutch smock mill with thatched cap and sides.
pollards *see* **middlings**

poll end large cast-iron socket on the end of the windshaft through which the stocks pass.
post the main upright post upon which a post mill revolves.
post mill this has a wooden top or buck, containing all the machinery, which is balanced upon a post and trestle. Post mills have no cap and the whole buck revolves round to face the wind.
quant four-sided shaft from stone nut to mace on overdriven stones.
quarter a mill, to to turn it slightly off the wind to slow it down.
quarter bars beams which reach diagonally from near the outer ends of the cross trees up to the sides of the post below the buck, taking the weight and bracing the whole trestle.
raddle red oxide used to smear checking gear to indicate raised areas on millstones.
rap block on feed shoe to bear against quant causing shoe to agitate.
reef, to to take in by rolling up or furling, some of the cloth of a common sail to reduce working surface.
rocking lever bar for controlling striking gear.
roller mill mill with grooved rollers. Not wind powered.
rollers bearings between curb and cap.
roundhouse the walled and roofed-in trestle part of a post mill, made to provide storage space for grain etc.
runner stone top stone of a pair which is turned by the mill.
sail backs *see* **whip**
sail bars bars attached to whips to carry cloth sails or shutters.
sails sweeps blown round by the wind to drive the machinery.
sailstocks beams passing through the canister to carry the whips.
scoop wheel vertical cast-iron wheel with wooden paddles or scoops which lifts the water from one level to another.
shaker arm arm attached to shoe which contacts damsel.
sharps *see* **middlings**
sheers main beams of cap frame.
shoe moving trough leading from hopper to the eye, which feeds the grain to the millstones.
shot curb curb and cap circle tracks with rollers between.
shutter bar bar which links spring-loaded shutters.
shutters movable vanes of springs and patent sails which open and close to present a working surface to the wind.
silk machine flour dresser with a silk sleeve.
sills horizontal timber plates on top of base walls of a smock mill to carry cant posts and framing.
sills, cap diagonal braces in cap frame.
skirt outer section of millstones.
smock mill wooden-framed, many-sided mill, clad with wood or thatch, with a movable cap.
spider iron cross on end of striking rod, linked to bell cranks and levers of striking gear.
spindle bar carrying stone nut which passes up through millstones to engage with the bridge and mace and carry the runner stone.
sprattle beam central beam across cap frame. The bearing of the upright shaft is on its underside.
spring-loaded shutters shutters of spring sails.
spring sail sail with shutters linked to a spring, the tension of which can be set manually so that the shutters will open and close according to wind strength.
steel proof steel bar against which the wood proof is checked for accuracy.
steelyard rod connecting governor to tentering mechanism.
stitching *see* **cracking**
stocks the long bars which cross through the top of the windshaft and carry the whips. Or, handles to hold mill bills.
stone dresser a man whose profession it is to re-sharpen or dress millstones.
stone nuts pinion wheels mounted on a spindle or quant which are moved into gear with the great spur wheel to drive the millstones.
striking rod rod which links spider to adjusting mechanism of patent sail by passing down through the hollow windshaft and out through its lower end.
supers *see* **middlings**
sweeps another term for sails.
tag attachment to the edge of the runner stone which sweeps the meal in the vat into chutes to bins below.
tail back end of a post mill.
tailpole long beam attached to the back of a mill by which it can be winded.
talthur small beam attached to tailpole which when hooked to the ladder will lift it clear of the ground while the mill is being winded.
tentering adjusting the gap between the millstones by raising or lowering the brayer and consequently the bridge tree which rests on it, carrying the lower end of the spindle.
tentering screw for adjusting distance between the stones by moving tentering mechanism by hand.
thrifts handles to hold mill bills.
thrust bearing any bearing taking sideways thrust of a shaft, as at the tail end of the windshaft.
tiver red ochre used for marking millstones. (*see* **raddle**)
tower mill mill built of brick or stone with movable cap.
tracer wooden staff used to check true movement of spindle.
trestle the whole of the support system of a post mill.
tun *see* **vat**
turret mill composite mill with round brick base carrying a curb on which the buck rests.
underdrift stone driven from below.
uplong longitudinal bar in a sail.
upright shaft *see* **main shaft**
vanes sails of a fantail or shutters of patent sails.
varying pitch twist from one end of the sail to the other. Sail bars are set on whips at a progressively greater angle from tip to heel of sail.
wallower mounted at the top of the upright driving shaft and meshing with the brake wheel. The first driven wheel of a mill.
warning bell bell which rings when the grain content of the hopper gets too low.
weather beam. *see* **breast beam**
weather studs upright bars in the weather beam which hold the neck bearing in position.
weathered sails sails with variable pitch from inner to outer end.
whips long bars attached to the front of the stocks which carry the sail bars, shutters etc.
winded turned to face the wind. Pronounced with a short 'i' to rhyme with 'sinned'.
winding gear tailpole or fantail for luffing into the wind.
windshaft axle on which the sails are mounted.
wip mill Dutch type of post mill. *see* **hollow post mill**
wire machine machine for separating flour from the rest of the meal.
wood proof wooden level for checking surface of millstones.

Acknowledgements

To write a book on windmills is to tackle a huge subject. There are enormous gaps in windmill records and a lot of work still has to be done. The work is only very loosely co-ordinated. Interested individuals and societies produce studies, books and papers on all kinds of technical aspects of the subject. I must acknowledge the fact that I have picked the brains of many of these people through their writings, while reserving my right sometimes to contradict their opinions. Except for one or two excellent books, so much of the work is technical and not very interesting to the uninitiated. I have tried to explain the basics in comprehensible terms and not to get bogged down in detail, for too much detail would be out of place here.

My thanks go to Mr Vincent Pargeter, millwright, for all his help in explaining technicalities, correcting my technical errors and for being there to argue with. He is a dedicated man with a deep knowledge of his craft and love of windmills. He is a perfectionist and a fine craftsman. That he is also a young man is fortunate for many windmills, which will in due course be restored by him. That he is also a good photographer and has spent many days travelling the country measuring and photographing mills has helped enormously in making this book, as his pictures have filled the gaps in my own collection.

Mr Anders Jesperson has been most kind in providing information and diagrams of Danish mills.

Maître André Darré, the miller of Coquelles, also helped me by allowing me to spend hours in his wonderful mill at Calais.

Thanks also to Mr Wootton, miller at Herne, for information about milling.

The Hilaire Belloc poem is reprinted by permission of A. D. Peters & Company.

Ginette Leach, who did most of the typing and helped with the research, deserves special thanks for her work on this book. She has carried camera bags, been patient while I and others have talked windmills for hours, got filthy clambering about inside old mills; while I have been thoroughly enjoying myself working on this book, which for me has not really been work at all, but a chance to study these wonderful things which caught my imagination as a child and have held it ever since.

PLATES

The plates in this volume are reproduced by kind permission of the following:

J. Bates: 106
Dutch Windmill Society: 63, 65, 66, 67
The Kent Messenger: 95
National Gallery, London: 5, 90
V. Pargeter: 2, 3b, 7, 13, 15, 23, 31, 35, 37, 41, 42, 47, 54, 61, 71, 72, 74, 75, 77, 78, 94, 97, 98, 99, 102, 103, 105, 107, 108, 109, 113, 114, 115
The Society for the Preservation of New England Antiquities: 109a and 109b
Photographs not listed above are from the author's collection

DIAGRAMS

Mr Bowles, Brighton Polytechnic: 24
Anders Jespersen, Mill Preservation Board, Denmark: 13, 14
V. Pargeter: 11a
Diagrams not listed above are by the author

Bibliography

BENNETT, Richard, and ELTON, John. *The History of Corn Milling*. Vol. 1: *Handstones, Slave and Cattle Mills*; Vol. 2: *Watermills and Windmills*; Vol 3: *Feudal Laws and Customs*; Vol 4: *Some Famous Feudal Mills* (London: Simpkin, Marshall, 1898 1901; Vol 2 repr by EP Publishing, 1973)

FREESE, Stanley. *Windmills and Millwrighting* (Newton Abbot: David & Charles, 1971)

Millnotes (published periodically by J. K. Major, 2 Eldon Road, Reading)

REYNOLDS, John. *Windmills and Watermills* (New York: Praeger, 1970)

STOCKHUYZEN, Frederick. *The Dutch Windmill* (Bussum, Holland: CAJ van Dishoek; English trans, London: Merlin Press, 1962)

VINCE, J. N. T. *Discovering Windmills* (Aylesbury: Shire Publication, 1969)

WAILES, Rex. *The English Windmill* (London: Routledge & Kegan Paul, 1954)

Index

Page numbers in *italics* refer to illustrations.